T0201882

GLIAL MAN

GLIAL MAN

A Revolution in Neuroscience

Yves Agid and Pierre Magistretti

Translated by

Robert N. Cory

Great Clarendon Street, Oxford, OX2 6DP,
United Kingdom

Oxford University Press is a department of the University of Oxford.
It furthers the University's objective of excellence in research, scholarship,
and education by publishing worldwide. Oxford is a registered trade mark of
Oxford University Press in the UK and in certain other countries

© Oxford University Press 2021

Illustrations by Yves Agid.
© ODILE JACOB, 2018
Originally published as *L'HOMME GLIAL: Une révolution dans les sciences du cerveau*,
Yves Agid et Pierre Magistretti, ODILE JACOB, 2018

The moral rights of the authors have been asserted

First Edition published in 2021

Impression: 1

All rights reserved. No part of this publication may be reproduced, stored in
a retrieval system, or transmitted, in any form or by any means, without the
prior permission in writing of Oxford University Press, or as expressly permitted
by law, by licence or under terms agreed with the appropriate reprographics
rights organization. Enquiries concerning reproduction outside the scope of the
above should be sent to the Rights Department, Oxford University Press, at the
address above

You must not circulate this work in any other form
and you must impose this same condition on any acquirer

Published in the United States of America by Oxford University Press
198 Madison Avenue, New York, NY 10016, United States of America

British Library Cataloguing in Publication Data

Data available

Library of Congress Control Number: 2020944416

ISBN 978–0–19–884767–0

Printed and bound by
CPI Group (UK) Ltd, Croydon, CR0 4YY

Oxford University Press makes no representation, express or implied, that the
drug dosages in this book are correct. Readers must therefore always check
the product information and clinical procedures with the most up-to-date
published product information and data sheets provided by the manufacturers
and the most recent codes of conduct and safety regulations. The authors and
the publishers do not accept responsibility or legal liability for any errors in the
text or for the misuse or misapplication of material in this work. Except where
otherwise stated, drug dosages and recommendations are for the non-pregnant
adult who is not breast-feeding

Links to third party websites are provided by Oxford in good faith and
for information only. Oxford disclaims any responsibility for the materials
contained in any third party website referenced in this work.

Acknowledgments

Yves Agid: this book was written, in large part, while I stayed abroad with friends in Martinique, first at the home of Claude Archambault (La Clémasol) and then in the company of Marie-Josée and John Persenda (Pointe Royale). I thank them wholeheartedly for their hospitality. Thanks also go to Agnès Renard who guided and supported me throughout this long period of writing.

Pierre Magistretti: many thanks to Christine, who has accompanied me throughout my 35-year glial journey.

Both authors: we are indebted to all those, who in one way or another, helped us see this book to completion; in particular we acknowledge the input of Andreas Hartmann, Jean-Léon Thomas, and Bernard Zalc. Thanks are also due Monica Navarro Suarez for compiling the references and, finally, our gratitude goes to Odile Jacob for editing and publishing the French edition of this manuscript and to Marie-Lorraine Colas for help in making it readable.

The translator of this English edition, Robert N. Cory, would like to express his gratitude to Mary Thaler, PhD, for her invaluable editorial assistance throughout this project; and to Martin Baum (Oxford University Press) and Shuichiro Takeda for their encouragement and support.

Contents

Introduction

Why do we need a different approach to understanding the human brain? And what do we mean when we use the phrase *glial man*? If we want to understand how systems function, whether mechanical or living, we need to be absolutely clear about their components. Presumably everyone knows what the brain is, and most have heard that it is composed of neurons. The word "neuron" has become such a part of our everyday vocabulary that a French politician, in response to a biting interview question, exclaimed: "give me credit for having at least two neurons to rub together!" It wouldn't have occurred to him to say ". . . two glial cells!" But it's those glial cells, those *non*-neuronal brain cells, that are the subject of this book.

Surprisingly, our brains contain more glial cells than neurons, and what's more, the proportion of glial cells to neurons has increased over evolutionary history (Herculano-Housel, 2009). For example, the nervous system of a leech contains six times more neurons than glial cells, whereas the highly evolved human brain has one and a half times more glial cells than neurons. It's even said that Einstein's brain contained more glial cells than the average human brain—could glial cells have contributed to his ingenuity?

In fact, there is much evidence to suggest this is so. Researchers who transplanted human glial cells into mouse brains found that the cognitive performance of the treated mice surpassed that of the control mice (Han et al., 2013). These observations, and many more, impel our interest in this neglected half of the brain.

Glial cells were first discovered in the mid-nineteenth century, but their importance went unnoticed until recently. It was the German physician Rudolph Virchow who first noted the existence of an amorphous glue-like substance interspersed among neurons. The term glia, meaning glue in Greek, was used to describe this interneuronal material. This "glue" was thought to serve as a support fabric, a type of brain connective tissue that held neurons in place by filling in the intervening empty space. It seemed like a logical, albeit rudimentary, explanation. With the development of new techniques affording us a closer look at glial cells and a better understanding of their inner workings, our perceptions have evolved. We've seen, for example, that glial cells respond to signals released by neurons, and that they, in turn, modulate neuronal activity. In some instances, their responses are even more precise than those of neurons. A particular type of glial cell, called the astrocyte due to its star-like shape (from the Greek word "aster," meaning star), is even required for memory consolidation! Not bad for a tissue once thought to be nothing more than inert connective matter.

What drove two scientists—one of us a neurologist, whose research solely involved neurons, and the other a neurobiologist, who accidentally came upon the existence of glial cells—to collaborate on this book? Well, despite our two very different—one might even say contrasting—career paths, we identified at least three reasons that suggest glial cells, that other half of the brain, are essential to understanding the complexities of behavior:

1. *The observation that glial cells are major participants in the communication between brain cells*, most notably communication with neurons—the cells most widely credited with creating thoughts. Today we can't imagine the functioning of the nervous system without considering these "Cinderella" cells. This is why the number of scientific articles appearing each year with the key word "glia" surged from 937 in 1985 to nearly 6,000 34 years later (Figure 1). We address this point in Chapter 1 of this book.

2. *The discovery that glial cells play an essential role in initiating and controlling our behaviors.* As shown later, their role is at least as indispensable as that of neurons.

3. *The realization that several nervous system diseases are tied to glial cell dysfunction* (see Chapter 4). The involvement of glial

cells in the disease process may be direct, i.e., the process starts with a glial cell anomaly that then affects neuronal activity; or their involvement may be indirect, i.e., a neuronal anomaly leads to glial cell dysfunction, which in turn aggravates the initial neuronal pathology, creating a vicious circle.

Figure 1. Number of articles on the subject of glia and neurons from 1985 to 2019

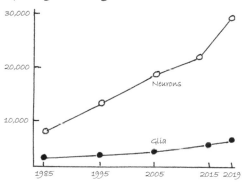

From 1985 to 2019, the number of articles on glia increased by nearly sixfold, but still remained well under the number of articles about neurons.

Today, glial cells are drawing increasing attention from the scientific community. Emerging from our "neuro-chauvinist" orientation, we have begun to recognize the importance of glial cells and especially astrocytes. But it takes time to accept new ideas. The psychologist William James noted three phases that accompany an innovative idea: Phase 1—it is incorrect; Phase 2—maybe it's true but it isn't important; and Phase 3—we always knew it was important. As far as glial cells go, we believe we are now in Phase 2. We can only hope Phase 3 arrives speedily.

Yves Agid—a career nearly devoid of glial cells!

"Over the past thirty years my team and I studied neurons, the brain's "celebrity" cells. We sought to identify neuronal abnormalities

underlying certain symptoms, as well as mechanisms responsible for the neuronal cell death observed in neurodegenerative disorders such as Alzheimer's and Parkinson's diseases. This productive research group saw several advances and significant discoveries. Curiously, over this long period of time there were few if any references to those other nervous system cells, the glia.

– Questioning: In the 1980s the term "gliosis" referred to an accumulation of glial cells in brain regions where neurons had died (for example, nerve tissue destroyed after the occlusion of an artery), but this term lacked scientific rigor. In other words, it was used for a clinical picture that had no clear explanation. The idea that an accumulation of glial cells "plugs up holes" left by the disappearance of neurons seemed just too mechanistic to me . . .

– Astonishment: I was invited to a scientific meeting organized by the Glial Cell Club that brought together most French researchers working in the field. I was surprised to learn that glial cells, those non-neuronal cells, display sophisticated biochemical properties and take on a major role in maintaining neurons.

– Revelation: In 2005, I stepped down as laboratory director and relinquished my administrative duties (grant applications, editing scientific publications, etc.). This gave me time to reflect. Having come to appreciate the role neural circuits play in orchestrating a variety of mental functions (memory and emotions, for example), I became more aware of glial cells, which make up more than half of the brain, and I seriously pondered their role in generating behavior, and even thought. Could it be that human thought is more than just neuronal? And if so, had Neuronal Man—*the book that inspired minds in the 1980s—evolved into* Glial Man? *I discussed this notion with my colleagues and we laughed. But I was captivated by the idea, more as a matter of principle than by a desire to rankle colleagues, but there was also my joyful inclination to take on cherished idols. One of my collaborators, Andreas Hartmann, began regularly sending me key scientific articles that appeared on the topic."*

Pierre Magistretti—chance and naiveté

"My interest in all things glial was sparked just over 35 years ago, but purely by chance. As a young medical graduate at Université de Genève I was headed into psychiatry. I thought to myself, if I really want to treat psychiatric patients, wouldn't it be better to first learn more about how the brain works? At that time, during my medical studies, the field of neuroscience was still in its infancy in Europe and so I decided to delve into the subject by pursuing graduate studies in the United States. In 1979, I began my PhD dissertation at the University of California, San Diego, under the guidance of Floyd Bloom, a neuroscience wunderkind at the time, whose lab focused on the biological bases of psychiatric disorders.

As soon as I arrived, Dr. Bloom asked me to develop a quick and simple biochemical test to assess the effects of the neurotransmitter norepinephrine, which seemed to play a role in depression. My research background was very limited, so I had no idea where to start. I recalled from my medical studies, however, that epinephrine was akin to norepinephrine, and that it can mobilize energy from the liver and muscle by breaking down glycogen—a stored form of glucose. Thus, I reasoned, norepinephrine must do likewise in the brain and I only had to measure the effects of norepinephrine on brain cells. To both my satisfaction and Floyd's, the experiments were conclusive and the lab adopted the test. This first success prompted me to dig deeper into the subject and I was really surprised to discover that brain glycogen is found primarily in one type of brain cell, the glial cells known as astrocytes.

I had a vague idea these cells existed and I remembered that they were supposed to be a kind of glue filling in the space between neurons. I was also unaware that the brain contains only low levels of glycogen. I must say, if I had known all this back then and had also been aware that glycogen is found mainly in that glue-like substance (Cali et al., 2016), I would certainly not have pursued work on that test!

Thus, beginning in total ignorance, but blessed with a measure of serendipity, I discovered that this glue is not so inert and in fact is key to comprehending how the brain functions. Since then, the field of neuron–glia interactions has grown by leaps and bounds and, as for me, I have worked in this area with the help of dozens of PhD students and researchers for over 40 years now!"

In this book, we seek to educate not only the public but also physicians and biologists on developments that could revolutionize our understanding of how the brain works, and we draw attention to these unique cells, for which new functions are regularly being discovered. There are three main types of glial cells in the nervous system; microglia, oligodendrocytes, and astrocytes:

1. The microglia don't have much in common with other glial cells; they are more akin to the blood cells called macrophages (see Appendix I).
2. The oligodendrocytes are glial cells that form the myelin sheath surrounding axons, like the insulation coating electrical wires. They are essential to proper neuronal functioning—myelin helps to accelerate the conduction of impulses along axons (see Appendix II).
3. Finally, there are astrocytes, the heroes of our story.

For simplicity's sake, and because less data is available on the first two cell types, this book is almost exclusively devoted to astrocytes, which we often just refer to as "glial cells."

So, the question to ponder is: will "glial man" topple Jean-Pierre Changeux's "neuronal man" (Changeux, 1983)? Are humans neuronal? Or are we glial? As we shall see, neither cell type can claim the crown—but to learn just where science has arrived on this question, we invite you to read the rest of this book!

1

The Brain

Neurons, Glial Cells, and Blood Vessels

Each one of us, naturally, has a personal stake in the human condition. Whether alone or as a community, we strive to survive and even better our lot. Humans are vulnerable creatures. We don't run very fast, nor are we particularly strong; and yet we dominate the animal world. In fact, no predator is fiercer—even with our own kind! Intelligent, skilled, and cunning; we possess the gift of spoken language; we know how to cooperate to defend ourselves; and we build and create. So, what gives us an edge over the rest of the animal kingdom? The answer lies in our powerful, exquisitely organized brain!

Our Brain Is in Charge of Our Body and of Itself

How does our brain provide us with this advantage? It is a wonderful machine, adept at filtering information from the environment. In response to the myriad signals arriving every fraction of a second, the brain generates behaviors that are amazingly adapted to our surroundings. Sitting astride the nervous system (Figure 2), it perceives what is happening outside the body through sensors located in the skin (touch) and other sensory organs (vision, hearing, smell, taste). The nerve signals generated by these sensors are transmitted through nerve pathways to the spinal cord, where they ascend to the brain. This perceptual information is processed by the posterior region of the brain and then relayed to its anterior region, the forebrain. The forebrain sends commands back down the spinal cord to

Figure 2. The brain oversees the body that carries it, as well as itself

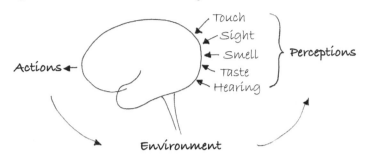

stimulate the nerves that activate our muscles, which then carry out the desired behaviors. What accounts for this nigh miraculous sequence of events? What mechanisms are involved?

The brain acts like a conductor who orchestrates the interaction between individuals and their environment. Its skillful direction relies on the proper functioning and organization of the neurons composing it. Three key principles influence the distribution of neurons in the brain:

1. *the development of the nervous system through the course of evolution (phylogenesis);*
2. *the posterior brain that perceives and the anterior brain that acts;*
3. *the specific layout of the neuronal regions* that oversee the three main modalities of human activity: motor skills, intellect, and emotions (see Appendix III).

But anatomical descriptions alone don't reveal how the brain works. For example, how does electrical information triggered by our sensory perceptions translate into action, language, and consciousness? How is this information integrated, synchronized, and coordinated to generate thought? The answers to these questions can be found in the distinct characteristics of the cells of the nervous system.

The prevailing dogma is that the brain's operations are solely "neuronal" and that they depend exclusively on the workings of neural circuits. Yet isn't it curious that the brain's functions, or dysfunctions—in the event of disease—are entirely attributed to neurons, despite the fact that they account for less than half of all brain cells?

The Brain, a Crowning Achievement!

To the naked eye the brain looks like a uniformly gelatinous mass. Not so—it is a work of art! It is also energy hungry, it's made of water and fat, it's protected, it's large, it's always in flux, it's smart, it's tenacious, it's intricate, and it's frequently misunderstood!

- *The brain is an energy hog.* Weighing in at about 1.3 kilograms (3 lbs.), only 2% of total body weight, it consumes 20% of the body's energy.
- *The brain is watery and fatty.* It is 75% water and the rest is mainly fat.
- *The brain is protected.* The blood–brain barrier, a relatively tight layer of cells within the brain's capillaries, protects the brain from toxic and immunological attacks.
- *The brain grows.* At birth the human brain is not yet mature. Its weight is 23% of an adult human brain—40% in the chimpanzee and 65% in the macaque. It continues to grow through adolescence, which explains why, unlike in other animals, the development of the human brain is highly dependent on how toddlers and children interact with the outside world. In other words, the environment and education are key to humans reaching their full potential.
- *The brain is in flux.* In adults it constantly evolves in response to learning. This is evident even when the brain is damaged, e.g., when a part of the brain is destroyed by the blockage of an artery (an ischemic stroke).
- *The brain is smart.* Each of its 85 billion neurons can adapt its activity and respond quickly. The brain is the commander and the one who perceives, decides, and acts.
- *The brain is tenacious.* Neurons do not die before we do—a 100-year-old human has century-old neurons.
- *The brain, like the universe, is an enigma that will be difficult to unravel.* One operates at an infinitely small level, the other on an infinitely large scale.

Neurons Are the Brain's "Thinking" Skeleton

Most people know that the brain is composed of neurons, but what are these neurons and what do they do? That turns out to be a long story.

Since the end of the nineteenth century the assumption that neurons are responsible for most brain functions has guided research. A now-obsolete notion was that the nervous system is a "syncytium"—more or less identical cellular elements connected to one another in a continuous web without intervening cell walls, much like a fisherman's net. This so-called reticular theory was expounded by the Italian histologist Camillo Golgi. Paradoxically, it was Golgi's very own cell staining technique (the eponymous Golgi stain) that led the Spanish histologist Santiago Ramón y Cajal to propose an alternative theory of brain functioning—the "neuron doctrine." According to Cajal, neurons are discrete cell units separated from one another and polarized, i.e., information flows from one end of the neuron (the dendrite) to the other (the axon). It's ironic that the same technical innovation, i.e., the Golgi stain, led Cajal and Golgi to reach such radically different conclusions (Ramón y Cajal, 1933). The two became fierce competitors and reportedly did not exchange a single word on the day they shared the Nobel Prize in 1906.

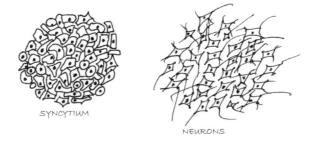

A fertile era of scientific inquiry ensued and numerous discoveries demonstrated that an electric current runs through the neuron. These findings fueled the golden age of neurophysiology, which continues to this day. As the 1950s rolled in, scientists showed that the "dry" neuron is actually "wet," i.e., it releases chemical substances and transmits signals from one unit to the next. Not long after, researchers started unravelling the neuronal basis for the major intellectual, emotional, and motor skill functions. These immense achievements in cognitive science were made possible by the advent of neuroimaging in humans.

Numerous models to explain how mental aptitudes function have been put forward; such as the global neuronal workspace model (see Appendix V), in which different brain regions communicate with one another in a coordinated and intelligent manner. Although these models derive from experimental observations, they are nonetheless still hypotheses. We understand that our brains "secrete" a lot of thought, but does this mean that neurons alone are responsible for our thoughts?

Connected Electric Wires?

It's hard to make out the shape of neurons—they exist in a densely woven mesh and are very different from the other cells in the body. Their cell bodies have a diameter of about 0.01 mm and contain a sphere-like nucleus. The neuron has several thousand extensions (processes) called dendrites—receptive zones that are in contact with the nerve endings of many surrounding neurons. A different extension, the axon, emerges from the cell body and terminates in tens of thousands of nerve endings that contact the receptive zones of neighboring neurons (Figure 3). Some axons exceed 1 m in length; for example, the neurons that originate in the spinal cord and project down to the feet. Others are shorter, less than 1 mm in length. Sometimes there are no axons at all, as is the case with interneurons, which play a role in the synchronization of neuronal activities (see Appendix IV).

Given that there are about 85 billion neurons in the human brain and that each one makes several tens of thousands of connections with its neighbors; and given that a neuron transmits around a thousand signals per second, it would seem we generate an astounding

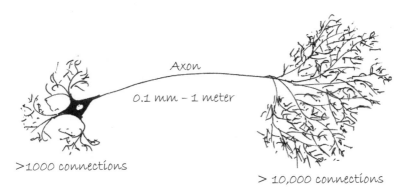

Figure 3. The human brain

☐ 1350 grams
☐ 85,000,000,000 neurons
☐ 1 mm³: 10–50,000 neurons

Axon

0.1 mm – 1 meter

>1000 connections

> 10,000 connections

one million trillion signals per second! The axonal nerve terminals of one neuron contact the dendrites of the next neuron in a zone called the synapse. The narrow gap in the synapse that separates the two neurons is 200 Å wide (or 0.00002 mm) and is called the synaptic cleft (Figure 4).

Most neurons establish strong connections between themselves—so-called point-to-point contacts. A neuron's nerve terminals secrete a chemical substance (a neurotransmitter) that is released into the

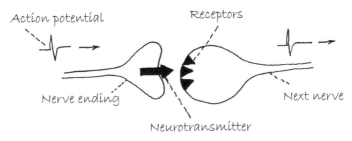

Figure 4. How do neurons transmit information?

Action potential

Receptors

Nerve ending

Next nerve

Neurotransmitter

Electrically (the action potential that runs along the axon) and chemically (the release of a neurotransmitter into the synapse)!

synapse and binds to receptors on the dendrites of the next neuron. Thus, direct communication is established between the two neurons, i.e., the activation of the first leads to the activation of the next.

By this mechanism, neurons keep messages moving throughout the brain. There are, however, a variety of neuron types—as many as several thousand. Most establish point-to-point contacts, but others are responsible for the harmonious function of the brain via neuromodulators (see Appendix IV).

Neurons relay their messages via an electric signal (the action potential) that propagates along the axon (Figure 4). Each action potential travels down the axon to the synapse where it triggers an activation or inhibition response in the neuron on the far side of the synaptic cleft. Apparently, our brain handles its information flow according to a binary law (activation or inhibition). It's always the same message sent the same way—via an action potential.

So, neurons form circuits that generate stereotypical responses using relatively fixed wiring. That's not to say neuronal signaling is totally inflexible; there remains some room for maneuver. Neurons can become more active or less active (more or fewer action potentials) and they have some capacity to rearrange themselves. In fact, neurons can move, though not in the usual sense of the word. Their collateral branches are able to migrate from one point to another.

Neurons can change over the short and the long term:

- *In the short term*, when you adapt to a sudden change, for example if you take off running, your neuronal electrical activity surges thanks to metabolic adaptations (e.g., the release of neurotransmitters into the synapse). This does not require any change in the neurons' anatomical architecture.
- *Over the long term*, for example when training for a sport, neural circuits are malleable enough to modify the distribution of their nerve connections in order to respond to your needs (Figure 5).

We must, however, distinguish between two different situations—an individual in a normal, healthy state versus someone in a

Figure 5. The brain learns and adapts because neurons are malleable

Electrical
stimulation

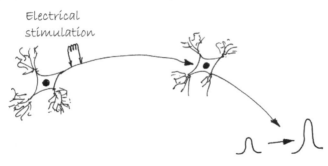

Repeated electrical stimulation of a neuron (for example, during learning) brings about a
heightened response in the downstream neuron.

pathological state. First take a normal situation in which learning is
required: groups of neurons modify their configuration by multi-
plying their synapses and, in doing so, become more efficient
at transmitting signals from one neuron to the next—so-called
synaptic plasticity (Ansermet and Magistretti, 2004). These new
connections form according to the demand and they ensure the efficient transfer of new information. As the information repeats, the
signals do not follow random pathways, but rather favor the same
pathways.

It's as if the newly acquired information develops the habit of
passing through newly remodeled synapses. These remodeled synapses are said to "strengthen." So, the acquisition of new information
creates new synaptic connections. Repeated activation of a neuron
can strengthen the downstream neural circuitry (long-term potentiation) or, conversely, weaken it (long-term depression). Thus,
repetitious learning, by engaging the same neural circuits over
and over, helps to create memories. It's as if, in our journey, several
paths lay before us at an intersection; but we always choose the same
well-trodden one.

The prevailing theory postulates that repetitive learning, which
gives rise to the phenomenon of memory, operates through synaptic
strengthening—thanks to the malleability of neural circuits (Figure 5).
But do these neurophysiological phenomena explain why we are able to
recall our high school graduation or our first day at work in our first job?

Figure 6. Neurons join forces when there is an injury

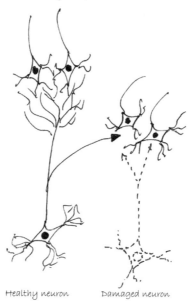

Healthy neuron   Damaged neuron

When a neuron is damaged and can no longer activate downstream neurons, healthy neurons in the vicinity send out collaterals to reactivate the disabled neurons.

Now consider a form of neuronal malleability associated with pathological situations. In the event of brain damage, for example the destruction of nervous tissue after a stroke, nerve endings mobilize to compensate for the destroyed neurons. It's like when a tree grows new branches to replace ones that were pruned. After the partial destruction of a group of neurons, the neighboring neurons that were spared have the capacity to fill the void left by the destroyed neurons (Figure 6). But this is true only up to a point, since the "branches" (nerve endings) can only grow back if the tree trunk (the cell body) has not been severed.

Complex Transportation Networks

Neurons may transmit information over long distances, but most often they organize into gigantic tangles of cell bodies and nerve endings in which it is difficult to discern how communication would function. Where does one start, given the nearly half a billion nerve connections per cubic millimeter of brain tissue? In fact, neuronal groups are organized in a perfectly ordered manner, like an immense network with major pathways (interstate highways) and tiny collaterals (rural byways). Here information is funneled and then sorted through vast "hubs," before being selectively dispatched to other brain regions. We don't completely understand the cybernetics of these brain communication pathways but, nonetheless, models abound to explain how they might function (see Appendix V).

My Brain Thinks, But Not My Computer ...

"My brain thinks, but not my computer," declared French philosopher André Comte-Sponville. And that's not all:

- *A computer's code is binary*, whereas the brain's is complex.
- *A computer's calculation is rapid, precise, irrevocable, and practically unlimited*; whereas the brain is slow and prone to mistakes. Also, it's apt to change its mind and has a limited capacity for calculation.
- *Computers communicate via series circuits*, whereas the brain works not only with series but also with parallel and recurring circuits.
- *Computers are not flexible*. In reality they have no intelligence aside from their programming and they are entirely obedient and apathetic. The brain, on the other hand, is impressively versatile; it can adapt, innovate, and motivate itself.

The nervous system consists of flexible neuronal networks that are highly branched. It does not operate by logic and calculation alone. It's ironic that we must rely on computers to help us make sense of the flow of information in this tangle of nerve cells.

Neurons Are Needed for Behavior

Without neurons we would have no behaviors. Witness what happens when a neural circuit is compromised or destroyed. The destruction or damage of a group of neurons may precipitate a variety of symptoms, including sensorimotor impairment (paralysis of half the body following a stroke or loss of sensitivity), intellectual deficits (aphasia caused by a lesion in the part of the cerebral cortex that controls language production, short-term memory loss caused by a lesion in a specific region called the hippocampus, etc.), or emotional disorders (disinhibition arising from damage to the neural pathways that control our emotions, various types of perseveration, depression, etc.). Conversely, the activation of a neural circuit can overstimulate motor behavior and disrupt motor skills (involuntary abnormal movements, such as tics), cause intellectual agitation (excitement, racing thoughts, etc.), or trigger emotional disorders (outbursts of laughter for no apparent reason, acute depression, etc.).

To understand the preeminent role attributed to neurons over the years, let's apply some straightforward logic. The brain produces thought and the brain is composed of neurons, so it naturally follows that neurons produce thought. And if neurons produce thought, they might also account for complex operations like the mental representation of a perception (a landscape, for example); or actions, like reading this book. In essence, the brain is a kind of neuronal machine that, by transmitting and manipulating symbols, allows us to modify our behaviors. The problem is that these symbols are ill-defined and the rules that govern their interrelationships are just now being deciphered.

Everyone agrees that neurons play a central role in the genesis of mental activities. This is true both for the brain's inputs, e.g., the inflow of visual information; and its outputs, e.g., the motor pathways that activate our muscles so they can express our behaviors. If neurons contribute to mental functions, does this also mean they generate very complex mental functions such as language or consciousness?

If neurons are merely channels of communication—which they are, but perhaps not exclusively—it follows that their interruption or activation would impact overall brain functioning. Imagine, as an analogy, that all the bridges across the Seine river in Paris were blown up and the main traffic arteries blocked. Traffic would come to a halt, store owners would close up shop, and the city's economy would be suddenly disrupted. But this doesn't necessarily mean the "intellectual" activity of the city would cease. Thought production, i.e., the source of the intelligence that drives the cultural, economic, and political life of the city, may well be generated elsewhere. Nonetheless, if Paris traffic were totally and permanently interrupted, the city's fate would be sealed in short order. If we apply this crude analogy to neuronal functioning, it seems safe to predict that the interruption in the flow of information within a neural circuit would result in behavioral impairment. But does this mean neurons think? What if the capacity for thought lay elsewhere?

Do Neurons Think?

If we could fully grasp all the electrical and chemical information arising from this incredible neuronal hodgepodge, would we attain a perfect understanding of the brain's operation? To this day most scientists will subscribe to the notion that the organization of neurons into circuits, of sometimes staggering complexity, constitutes the template for the production of all sophisticated mental functions. We really should critically evaluate this point of view. Despite the conceptual niceties of the models proposed along these lines, all attempts to validate them have proven inconclusive. More to the point, it's astonishing that the anatomical organization of neural circuits appears to be so immutable and the neuron's biological mechanisms so stereotyped and routine. Is it possible that mere neuron-to-neuron connections can explain the vast complexity of our behaviors, or even simple ones like writing with a pen, recalling a scene from the past, or anticipating an action?

There is no doubt that neurons are essential to producing behaviors. We know that when we activate a neural circuit we trigger behaviors and, if we destroy a given group of neurons, we may suppress those same behaviors. Nonetheless, despite more than a century of research in the field of neurophysiology, scientists have yet to prove a causal relationship between synaptic activity and specific mental activities—though one may indeed exist.

What Happens When I Take Up My Pen?

My pen sits on the table and I perceive it. I see its shape, position, and color. In the back of my eyes, my retinas receive this visual information and miraculously transform the pen's image into an electric current that runs through a series of neural circuits until it reaches the posterior region of my cerebral cortex, an area that integrates visual information. At that point I recognize this is indeed a pen; it possesses all the basic features of a pen. It is, in fact, a specific pen distinguished by its shape, color, and brand. How do I recognize that this is *my* pen? I connect its image with a memory catalog of objects stored in my brain, including the memory of this specific variety of pen. How does this information, which traverses my brain in the form of small electric impulses, turn into the mental representation of a pen that allows me to recognize it with certainty? We don't yet have the whole answer to this question. How do photons (which bring the pen's image to my retina), then electric impulses (which convey the pen object information to various parts of my brain) lead to the realization that this is a pen? In other words, how do physical and electrical data give rise to a psychological process as complex as the perception and recognition of an object? For now, there is a gap in our knowledge of how the simple perception of an object relates to a complex psychological process.

Next, I start using my pen. When I get ready to write, the neural information created by the recognition of the pen in the posterior part of my brain is forwarded to the anterior part, namely the frontal cortex. The latter orchestrates a complex motor behavior that enables me to manipulate my pen and write. To complete this complex task the information from my frontal cortex is conveyed, via my spinal cord and peripheral nerves, down to the muscles of my right arm and hand (Agid, 2013).

And so, the act of writing commences.

Are Neurons Nothing More Than Brain Communication Pathways?

If neurons are indeed instruments of thought, can we assert, as well, that science has worked out the circuits underlying thought transmission? Yes, but in a rudimentary way. Neurons function like avenues when they assemble into major neural tracts, as streets when they gather together into neural pathways, and as alleyways when they form miniscule nerve fibers. Returning to the Paris traffic analogy, how would this city look to the inhabitants of a faraway planet (Agid, 2013) (Figure 7)? They would notice not only the streets, but also the buildings on either side of these streets.

Imagine these extraterrestrial beings gazing at Earth through a telescope so powerful that they could make out the tiniest details. They zoom in on the Champs Élysées and, in the midst of the traffic congestion, identify a delivery truck. They notice that every morning this truck begins its trip at the same building and follows the same meandering route to deliver its parcels.

One of these far-off earth-gazers is also a diligent scientist and jots down this observation in a notebook: "In a city on another planet I discovered the existence of a mobile object that at a set time everyday follows the same route.

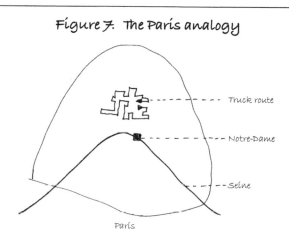

Figure 7. The Paris analogy

Truck route

Notre-Dame

Seine

Paris

It's perfectly reasonable for a distant observer to wonder if knowing the complex dynamics underlying the movement of vehicles along Parisian streets could shed light on the city's overall operation. Is it possible that just by comprehending these traffic patterns one can figure out the city's inner workings? Suppose instead the city were governed from within the buildings that line the thoroughfares?

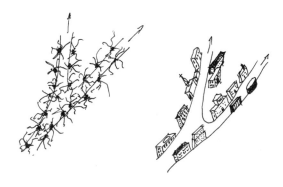

> Now let's transpose these imaginary observations to the brain's func-
> tioning. The streets that provide the means of communication are the
> neurons, while the neighboring houses are the astrocytes. Might it be that
> the processes governing our behavior emerge from within these astro-
> cytes? And could our thoughts, and even our thoughts *about* thought, by
> which we mean our consciousness, emerge from these same cells? But now
> we are truly taking flight—let's bring our speculations back down to Earth!

This trip attests to a specific dynamic that I can measure by following a marker, i.e., a vehicle. Since I have confirmed this observation by examining other vehicles following other routes, I draw two
conclusions:

1. *There are sophisticated communication pathways in this city*
   along which a phenomenal amount of information is conveyed. It seems these communication pathways may have the
   capacity to learn. If this were true, these pathways would support an actual memory function. This is a logical conclusion,
   but I wonder how the operation of these diverse communication pathways can account for the political, economic, and cultural operations of this city.
2. *These communication pathways do not exist in isolation.* They
   are lined with buildings on either side. What could be going
   on inside these buildings? Could the decisions made in these
   buildings govern the city's operations? My telescope doesn't
   help me answer these questions. To understand, I need to find
   out what takes place inside those buildings."

So how do neurons, with their more-or-less fixed wiring and binary stereotyped responses (activation or inhibition), manage to
produce the phenomenon of thought? Can this be explained by the
number of neurons and their collateral branches? What accounts
for the specificity of neural messages, given that neurons receive
a vast array of information from diverse sources? What happens

when new information enters the queue, or when signals are coordinated incorrectly? And, an even more challenging question, how can the neural code be deciphered (Dehaene, 2014)? How does a pattern of electrochemical currents flowing through a neural network engender a mental representation? Something, evidently, it

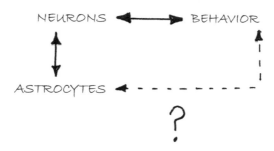

does. It's as if the neural circuitry must attain a high level of complexity before thoughts can be produced, i.e., brain matter generates thought only after a certain threshold of spatial and temporal organization is surpassed. So far, so good. But the difficulty is that, for the time being, there's no conclusive evidence to support this conjecture. Irrespective of the elegance of the neural code paradigm, attempts to demonstrate its validity have fallen short. So why not entertain some unorthodox notions at this point, such as, for example, the possibility that glial cells play a role in the generation of thought!

Let's articulate the appropriate question: How would glial cells, more specifically astrocytes, participate in generating thought? In particular, do astrocytes contribute to the development of behaviors? One possibility is that astrocytes act through neurons that, in turn, determine behaviors. Another idea, a totally heretical one, is that astrocytes are directly responsible for behavior (Robertson, 2002)! For now, let's start with the very reasonable proposition that our nerves convey environmental perceptions to the brain via sensory neuron pathways and produce actions via motor neuron pathways—the latter originate in the cerebral cortex and descend the spinal cord to reach the muscles. In sum, neurons are responsible for transmitting all information via incoming (afferent) and outgoing (efferent) pathways.

But what transpires inside the brain? Are the brain's neurons alone responsible for processing and interpreting all afferent information? Do these neurons alone cause us to ponder, dream, and contemplate? Given that neurons are intimately associated with astrocytes, with which they interact constantly, perhaps the neuron–astrocyte pair is actually the basis for all mental processing? What's that you say? Astrocytes—those mere foot soldiers who serve the neuronal top brass—may actually be involved in mental operations? Now that's an intriguing thought!

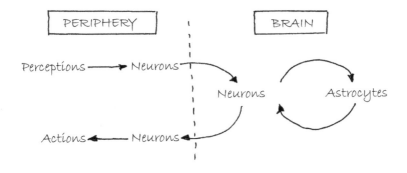

To recap, it strains belief to say that the brain's normal operations— or in the event of disease, its abnormal ones—can be solely attributed to neurons, which in fact account for less than 50% of the brain's cells. Indeed, there are many non-neuronal brain cells! So, what are these other brain cells up to?

Why Have Glial Cells Fallen Off the Radar?

When someone seems dimwitted, we might remark, "they are missing a few neurons (those proverbial marbles)." We would never say, "they are missing a few glial cells." Even the well-educated among us may not know the word "glia." But for biologists it's a different story, especially those in the field of neuroscience. The latter fall into two categories: glial cell specialists, who contribute to an international scientific journal Glia, and the more numerous neuronal specialists, who also have their own specialty journal called Neuron. Wouldn't it be nice if these two camps actually communicated more often!

Neurons play a vital role conveying the impulses required for mental operations, but what are glial cells up to? Are they merely custodial staff attending to neurons—feeding them and carrying away their waste? Imagine the opposite were true. Imagine neurons were there to support glial cells. Is this suggestion too provocative to be taken seriously? Not at all. The science of neurons has made great strides in recent years; but this is even more true for glial research, and especially the study of astrocytes, for which our knowledge has grown by leaps and bounds in the past two decades. Numerous review articles have been published in this field (Nedergaard et al., 2003; Miller, 2005; Volterra and Meldolesi, 2005; Halassa et al., 2007; Zhang and Ben Barres, 2010; Eroglu and Ben Barres, 2010; Pannasch and Rouach, 2013; Verkhratsky and Butt, 2013 ; Elsayed and Magistretti, 2015; Haydon and Nedergaard, 2015; Magistretti and Allaman, 2015).

More Complex and Numerous, and yet Overlooked . . .

There are about 85 billion neurons in the human brain and somewhat more glial cells—over 100 billion. These are approximate numbers; glial cells may outnumber neurons by anywhere from 120% to 160% (Herculano-Houzel, 2009). The percentage of glial cells relative to neurons varies by brain region (from 20% in the cerebellum up to 160% in the cerebral cortex) and by species (55% in mice vs. 120% in humans). We must also bear in mind the many oligodendrocytes that make up myelin (Appendix II). Myelin is the primary constituent of white matter, and its contribution to brain volume amounts to 12% in mice and an impressive 55% in humans (Zalc and Rosier, 2018).

As humans evolved, the number and complexity of glial cells outpaced neurons. The ratio of glial cells to neurons also increased (Figure 8). This could be interpreted in one of two ways: either glial cells are key to enabling major mental functions and neurons developed merely to support glial dominance; or glial cells evolved in response to the growing complexity of neurons, which retain their role as the predominant makers of thought.

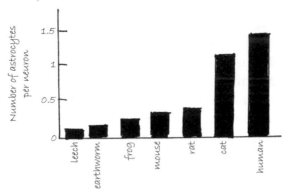

Figure 8. Astrocytes: a cognitive index of evolution?

As evolution progressed, the brain expanded, as did the number of astrocytes. A slug, with its ganglion of only 25 to 30 neurons, has just a single astrocyte. The earthworm has six times as many neurons as glial cells. In flies, rats, and humans the percentage of glial cells relative to neurons is 20%, 60%, and 120%, respectively. Throughout evolution the number of glial cells increased more rapidly than the number of neurons as the nervous system became more powerful.

How to explain this? These differences cannot be attributed to differences in metabolism, which doesn't vary much across species, especially in vertebrates. There are two possible answers. It could be that the increase in the number of astrocytes during evolution is tied to the increasing complexity of the neural network. It's a known fact that the more complex and denser this network became, the greater were its nutritional needs, which in turn required more astrocytes to handle the surging demand. However, the opposite could also be true, i.e. the increase in the number of nerve endings and synaptic contacts was a consequence of increased

astrocyte density. We know astrocytes became larger and more sophisticated through evolution. This latter hypothesis assumes that the large
number of glial cells relative to neurons was the main factor driving the expansion of mental functions.

The higher relative glial cell density observed in a post-mortem study
on the brain of Albert Einstein (Diamond, 1985) seems highly suggestive,
though, of course, we can't draw conclusions from just a single case.

EINSTEIN

Furthermore, as evolution proceeded, neuronal density (the ratio
of neuron number to brain volume) decreased as the overall volume
of nerve tissue increased! At the same time, the density of glial cells
stayed relatively constant, varying little between brain regions and
across species. It's tempting to conclude that the increasing ratio of
glial cells to neurons during evolution accounts for the growth in
our intelligence. But a word of caution here. Although the density
of neurons decreased during evolution, the density of nerve endings
remained relatively stable (Herculano-Houzel, 2014). This makes it
hard to draw any firm conclusions.

Within the brain of any one species, the size of neurons (including
the cell body and the volume occupied by the dendrite branches)
may vary by up to 160-fold, depending on the brain region; whereas
the size of glial cells will vary by only 1.4-fold. In the course of evolution, the shape of neurons became more complex (cell body
plus dendritic branching); but this occurred to a lesser extent with
glial cells.

A human astrocyte is three times the diameter and structurally more complex than a rodent astrocyte. The human astrocyte also has ten times the number of processes. Thus, a human astrocyte occupies 16 times the volume of a rodent astrocyte (Oberheim et al., 2009). The relative increase in the number of glial cells and their processes compared to neurons can be explained by the proliferation of neuronal nerve endings, which generate additional nutritional requirements due to their high level of metabolism. Many glial cells are needed to tend to all these synaptic terminals. It appears that in humans, a single human astrocyte may occupy an area of the brain "covering" up to two million synapses, compared to just 120,000 in rodents (Oberheim et al., 2006, 2009).

So, the overall cellular architecture of astrocytes became more intricate, even though their key morphological features do not vary much across species—for example, between flies and humans (Figure 9)—suggesting that the major functions of astrocytes have been preserved through evolution. The development of this more elaborate glial cell morphology over the course of evolution may have played a role in mental development. But we must be careful about resorting to differences in phenotypes (observable traits and characteristics) to explain differences in intellectual abilities among humans! What we can safely assert is that in humans the number and complexity of astrocytes has grown as the brain has become more complex.

Figure 9. Size of an astrocyte in a fly and in humans

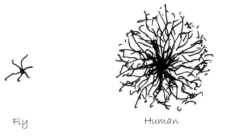

Fly   Human

The Intriguing Discovery of Glial Cells

The discovery of glial cells is attributed to the German neuropathologist Rudolf Virchow in 1856 (Somjen, 1988; Kettenmann and Verkhratsky, 2008). Virchow didn't describe glial cells as such; he wrote about a non-specific biological connective tissue that filled the space between nerve cells and separated them from blood vessels. He called this cement-like tissue that held cells together the "nerve glue," or Nervenkitt in German. By 1865 the German neuroanatomist Otto Deiters had identified non-neuronal cells. The end of the nineteenth century and the start of the twentieth saw a fertile period of scientific discovery during which scientists used rudimentary microscopes to elucidate the properties of these cells that would later be called glial cells.

Things really took off with Camillo Golgi, whose silver staining technique distinguished the neurons, with their characteristic axons, from cell types without axons (Golgi, 1873). In addition, he demonstrated the relationship between glial cells and the neighboring blood vessels. He hypothesized, along with his student Sala, that glial cells play a role in transporting nutrients out of the circulatory system to the neurons. A few years later Andriezen (1893) confirmed this hypothesis.

Ramón y Cajal (1913) speculated on the function of these new cellular elements. He was ahead of his time because, as he himself admitted, neuroscientists did not yet possess the means to pin down the function of these glial cells. He nevertheless observed that glial cells strongly resemble endocrine cells, due to the presence of secretory granules. He suggested that glial cells are not only involved in feeding neurons but that they also play a role in waste elimination, a concept advanced by Marinesco (1896) and Lugaro (1907). In addition, he postulated that embryonic glial cells are responsible for the migration of neurons. Building on mere histological descriptions, Cajal was able to elaborate a veritable physiological theory of brain functioning at the cellular level (Figure 10)!

Figure 10. Two astrocytes in the human cerebral cortex, one "embraces" a pyramidal neuron (Araque et al., 2014)

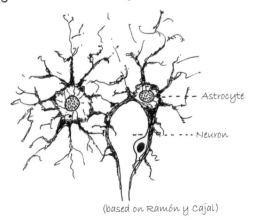

(based on Ramón y Cajal)

Curiously, it was only later that Cajal's student Del Rio Ortega (1920) introduced the term "microglia" to describe a new class of non-neuronal nerve cells (Appendix I) and it is to him we owe the first description of oligodendrocytes, the glial cells responsible for the myelin sheath that wraps around neuronal axons.

WHAT COULD NEURONS POSSIBLY BE UP TO?

Glial cells, therefore, are not just inert cement that holds neurons together within brain tissue. It would take nearly 100 years of scientific research to demonstrate that astrocytes: (1) ensure an energy supply so that neurons are able to function optimally, (2) participate in the communication between all nerve cells, (3) control the formation of synapses and the creation of new neurons, and (4) play a role in generating behavior.

Every Shape of Astrocyte in Its Place

As previously mentioned, astrocytes get their name from their star-like shape (in Greek astros means "star") (Figure 11). Two basic features clue us in to their physiological role:

Figure 11. Three types of astrocytes

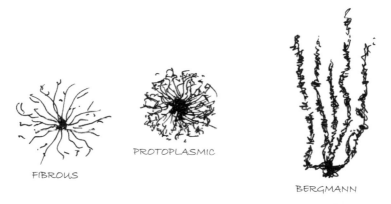

FIBROUS

PROTOPLASMIC

BERGMANN

Fibrous astrocytes are distributed throughout myelin, protoplasmic astrocytes are found in the gray matter (where the neurons are), and Bergmann cells are located in the cerebellum.

1. *Astrocyte processes are in contact with blood vessels*, which provide the brain with energy in the form of glucose and oxygen. Also, these processes surround most synapses, so a single astrocyte may envelop several hundred thousand synapses belonging to a hundred or so neurons. Through this arrangement astrocytes monitor the synaptic activity level and couple it to the brain's energy uptake.

2. *Astrocytes, with their regular polyhedral shapes, fill space in an orderly fashion*—only about 5% of their volume overlaps with their neighbors. This type of spatial organization has clear physiological implications, in contrast to the arrangement of neurons that, at first glance, appears chaotic (Nedergaard et al., 2003).

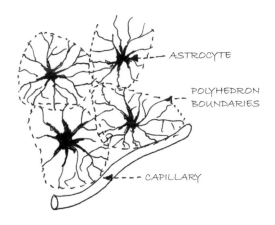

Blood Vessels, the Third Component

Intricate Plumbing

The brain's blood vessels receive less attention, yet they are clearly essential to the brain's normal functioning. They deliver nourishment, carry away waste, and form a barrier between the blood and the brain, which they insulate and protect. Molecules cross this blood–brain barrier selectively, either via specific carriers—as is the case with glucose, an essential brain nutrient—or on account of their biophysical properties, i.e., lipophilic molecules are more permeable and cross the lipid-based cell membranes more readily than do hydrophilic molecules. Blood vessels continuously supply the brain, adjusting the blood flow to meet its ever-changing requirements.

This plumbing network is sophisticated yet vulnerable. When things go wrong, for example during a respiratory crisis that reduces the brain's oxygen supply, nerve cells become stressed within seconds and die after a few minutes. It doesn't take a scientific whiz to see that in the event of neurodegenerative disease, any blood supply alteration will exacerbate the pathological process. A vascular disruption can result from this process and then, in turn, accelerate the disease in a sort of vicious cycle, or it may actually be the root cause of the disease process (discussed further in Chapter 4).

Astrocytes surround blood capillaries and cling to their walls with a large number of pseudopodia (Takano et al., 2006). These play a two-fold role; they assist with nutrient supply and waste disposal, and they protect neurons from external attacks.

Figure 12. The blood vessels

20 microns

Blood vessels don't think, but they are indispensable to thought.

The blood irrigates the brain via four large arteries (the right and left carotid arteries at the front of the neck, and the right and left vertebral arteries at the rear of the neck). These arteries subdivide into smaller arterioles that supply oxygenated blood, the brain's fuel. Similarly, waste is collected into small venules and carried away by the large veins.

Take a moment to consider some impressive numbers. Blood vessels account for about 5% of the brain's volume. Placed end to end, they would stretch over 600 km (Weber et al., 2008)! They are in contact with the blood over a surface area of about 20 m². The three-dimensional complexity of this capillary organization is such that it seems every neuron in the brain may have its own capillary.

An Efficient Gatekeeper

The blood–brain barrier (Figure 13) controls the exchange of molecules between the brain and the rest of the body, a necessity for the brain's normal development and functioning. It's a true physical barrier composed of two cell layers that make up the interface between the circulating blood and the brain tissue. Some of these cells, the endothelial cells, are in contact with the blood and form the walls of the hair-sized blood vessels called capillaries. The other cells, the astrocytes, face in toward brain, but at the same time interact directly with the endothelial cells (Nedergaard et al., 2003).

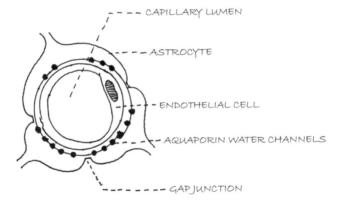

Figure 13. The blood–brain barrier

- - - - CAPILLARY LUMEN

- - - - ASTROCYTE

- - ENDOTHELIAL CELL

- - AQUAPORIN WATER CHANNELS

- - - - - - GAP JUNCTION

The blood–brain barrier is a dynamic filter protecting the brain and insulating it from the rest of the body. For example, when there is an infection in the body, the microbes cannot, in principle, pass from the blood into the brain. The two cell layers, the endothelial cells that form the vascular wall and the brain astrocytes, prevent or markedly reduce the flow of molecules passing from the blood into the brain and vice versa.

Large molecules such as proteins, or cells such as red blood cells, cannot cross. In addition, 98% of all small molecules cannot pass, nor can most drugs. The few small molecules that are able to cross the blood–brain barrier include amino acids, which are conveyed via specific carriers; and others with specific biophysical properties, among them vitamins and certain drugs. One might say that the blood–brain barrier serves as the brain's smart border guard.

Neurons, Astrocytes, and Blood Vessels: A Veritable "*Ménage à Trois*"

Astrocytes, neurons, and blood vessels are engaged in an intimate relationship. At the center of this arrangement are the astrocytes and their processes that extend to both neurons and capillaries. These astrocytic "feet" reach out to the bloodstream and let the astrocytes control the blood flow and the traffic of molecules to and from the brain.

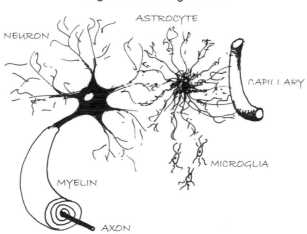

Figure 14. Ménage à trois

The regulation of blood flow and molecular traffic is in large part coupled to neuronal activity, which the astrocytes monitor via their many processes that surround the synapses. Astrocytes play a central role coordinating the *ménage à trois* by maintaining a functional relationship between the capillaries and the neurons (Figure 14). The question is: how do they do it?

2
Astrocytes

A Key Player in Brain Functions

Ingenious technological developments in the early 1990s made it possible to visualize brain cell activity, bringing initial insights into the neuron–glia dialogue. Up to that point, the only way to detect the activity of neurons was to record their electrical signals with electrodes. The electrodes were usually coupled to amplifiers, so one could literally hear the activity of a single neuron, or at most several dozen neurons simultaneously. However, unlike neurons, astrocytes produce no electrical signals, and so remained silent.

Our ability to detect astrocytic activity followed advancements in microscopy and the discovery of a cell activity marker. This marker, an increase in intracellular calcium levels, works with any cell type, including astrocytes (Cornell-Bell et al., 1990). Many neurotransmitter receptors promote the uptake of calcium by cells and, in some cases, also mobilize the calcium stored within intracellular compartments. The result is a surge in intracellular calcium levels that activates a variety of intracellular signaling pathways (Clapham, 1995).

Using new microscopic techniques and calcium-activated fluorescent markers, we are now able to visualize the presence of calcium with a high degree of spatial and temporal accuracy. These technical advances have brought about a revolution in the study of astrocytes—*And after darkness, light!* (Job 17:12)

We are now aware that a number of astrocytic receptors, in particular those activated by glutamate, trigger a rapid increase in intracellular calcium levels. Previously we had no means to record or listen to this activity, but we can now see it and it turns out that these supposedly silent cells put on quite a fireworks display!

Figure 15. The astrocyte tsunami

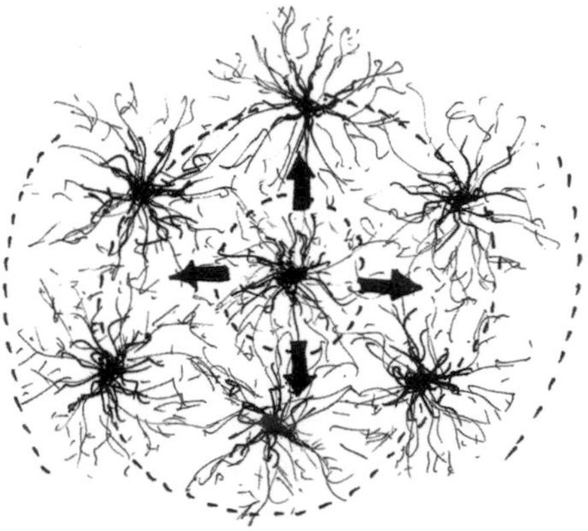

Waves of calcium (arrows) spread out over long distances (several millimeters) and wash over neighboring astrocytes.

The fluorescent signals linked to calcium release are triggered by glutamate and are not limited to a single astrocyte. Instead, the calcium propagates from astrocyte to astrocyte in the form of "waves" that fan out like a tsunami (Charles, 1998) (Figure 15).

Telephone Versus Radio: How Do We Communicate?

The function of astrocytic calcium waves is still hotly debated, but it is consistent with another unique feature of astrocytes: the existence of intercellular tunnels that connect each astrocyte to scores of other astrocytes, thereby creating an astrocytic network—a veritable "super cell", or syncytium. These tunnels, called gap junctions, are made up of proteins that create pores in the cell membrane that allow ions (for example, calcium) or small molecules such as glucose or lactate to pass from one astrocyte to the next (Theis and Giaume, 2012) (Figure 16).

Thus, in the nervous system there are two ways for cells to communicate with one another: either via synapses between neurons or

Figure 16. Astrocytes communicate with one another via tunnels called gap junctions

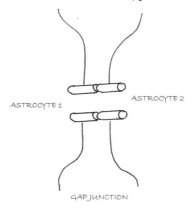

Astrocytes are connected to each other by gap junctions. These junctions are composed of six transmembrane protein units called connexins. Together these proteins form channels through which astrocytes exchange ions and small molecules (up to 1,200 Daltons in molecular weight).

via gap junctions between astrocytes. The first method is more precise and is limited to point-to-point communication between two neurons—we could compare this to the communication between two telephone receivers. The second method, via gap junctions between astrocytes, is more similar to radio communication—a signal emitted by a transmitter reaches hundreds of radio receivers simultaneously. But these two modes of communication are interrelated; in addition to their synapses, neurons can communicate directly with astrocytes, provided the latter have receptors for certain neurotransmitters. For example, glutamate is released into a synapse by Neuron A and then Neuron B is activated (Figure 17). The synapse between A and B is surrounded by astrocytic processes that also have glutamate receptors. When these astrocytic receptors are activated, they trigger calcium waves that spread to hundreds of other astrocytes. The signal that was initially limited to Neurons A and B is now amplified and retransmitted over a much broader range.

Figure 17: "Telephone" communication versus "radio" communication

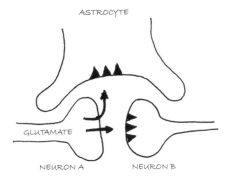

Communication at the synaptic level proceeds in a precise manner between Neuron A and Neuron B, like between two telephone devices. On the other hand, astrocyte-mediated communication reaches multiple synapses at once, comparable to a radio transmission broadcast to many listeners at a time.

The question remains: what is the purpose of this hybrid communication? Nature tends to be parsimonious and despite current gaps in our understanding, it would be surprising if this mechanism were not somehow involved in the brain's information processing functions (Volterra and Meldolesi, 2005).

Astrocytes Multitask: They Supply Food and Carry Off Waste

The brain consumes 20% of the body's energy, but where does all this energy go? To the neurons. And who supplies this energy to the neurons? The astrocytes. The latter provide neurons with all the nourishment they need at any given moment. Astrocytes monitor neuronal activity, for example, by detecting their metabolic activity and the calcium waves surrounding them. In response to the neurons' increased energy demands, astrocytes signal blood vessels, through their processes, to vasodilate, which allows more glucose to reach active brain regions (Zonta et al., 2003; Howarth, 2014) (Figure 18).

The greater the increase in neuronal activity in any given brain region, the more glucose the astrocytes pump into that area. However, neurons do not consume glucose directly. Their preferred food is lactate, a product made from glucose (Magistretti and Allaman, 2015, 2018). Glucose is metabolized inside astrocytes in a sort of "pre-digestion" step that produces lactate, which is then "fed" to neurons. Neurons use this lactate molecule to generate about 15 molecules of ATP (adenosine triphosphate), the fuel used by every cell in the body (Pellerin and Magistretti, 1997; Machler et al., 2016).

Besides feeding neurons, astrocytes also clean up after them. They eliminate and recycle compounds produced by neurons, including neurotransmitters, the chemicals that mediate the transfer of information from one neuron to the next (Gundersen et al., 2015). How do astrocytes accomplish this? It requires more than just being in the right place at the right time. We have just seen that astrocytes have molecular mechanisms that detect neuronal activity at the synaptic level and facilitate energy supply to the brain. It turns out that astrocytes have receptors and uptake sites that recognize the neurotransmitters released into synapses, especially glutamate—the most common signal of neuronal activity, which is released by 80% of all synapses (Allaman et al., 2011; Gundersen et al., 2015). When glutamate molecules are released into synapses, they transmit information to other neurons and then are quickly scooped up by astrocytes.

There are two reasons why astrocytes must eliminate neurotransmitters such as glutamate (Figure 18). First, glutamate is the main agent that neurons use to communicate across synapses. If glutamate were released into a synapse already flooded with glutamate, the poor signal-to-noise ratio would compromise interneuronal communication. It would be like dropping a bucket of water into the ocean—there would be no observable effect. But this same bucket of water means a great deal to a thirsty plant! The astrocyte's job is, literally, to suck the glutamate out of the synapse once the neurotransmitter has relayed its signal. In doing so, it maintains the optimal signal-to-noise ratio needed for proper neurotransmission (see Figure 18). The second reason for the elimination of synaptic glutamate by astrocytes is that excess glutamate endangers neurons—a phenomenon called excitotoxicity (Olney, 1969; Sattler and Tymianski, 2001). Glutamate acts upon certain receptors that stimulate

Figure 18. Glucose, lactate, and glutamate

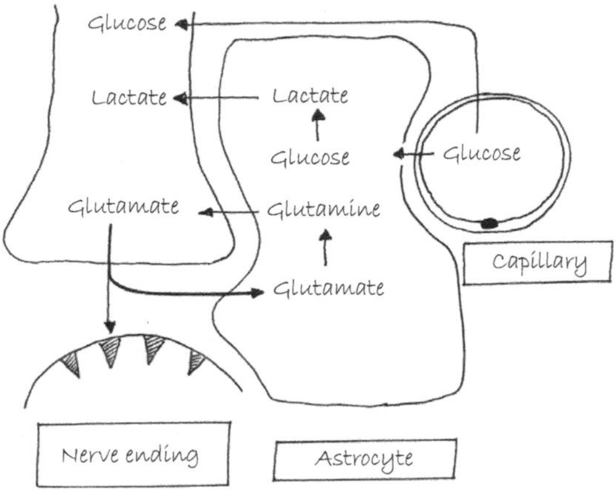

The glutamate released by neurons generates a signal in astrocytes. They respond by importing glucose from the bloodstream, transforming the glucose into lactate, and delivering it to the neurons (Pellerin and Magistretti, 1994).

the entry of calcium, which at moderate levels is a normal cellular signal. But excess glutamate overactivates receptors and leads to high calcium concentrations, which triggers a pathological process that can kill nerve cells. By quickly clearing glutamate from the synapse, astrocytes guard against this excitotoxicity threat.

What do the astrocytes do with the glutamate they've taken up? Each glutamate molecule enters the astrocyte via a carrier, a sort of molecular channel that enables substances to cross the plasma membrane. During its transport the glutamate is accompanied by three sodium ions. Bringing in both glutamate and sodium is no easy job for the astrocyte—it's a major metabolic task! As with any cell, the astrocyte's intracellular sodium levels are kept relatively constant. When sodium enters the cell, regardless of the means, the astrocyte must remove it to maintain a steady intracellular concentration of this ion. This task falls to an enzyme called sodium–potassium ATPase, which consumes energy. To eliminate three sodium ions, this enzyme must consume one molecule of ATP. Once inside the astrocyte, glutamate is converted into the amino acid glutamine, which is then re-routed back to the neurons (the neurons convert the glutamine back into glutamate). However, the transformation of glutamate into glutamine within the astrocyte also costs one molecule of ATP. So, every time an astrocyte takes up one molecule of glutamate from the synapse, it must pay a two-ATP-molecule tax—one ATP to eliminate the sodium and one to convert glutamate into glutamine (Magistretti and Chatton, 2005). Once the ATP is consumed, its intracellular concentration falls and the astrocytes must take up glucose in order to produce more ATP (Voutsinos-Porche et al., 2003).

Astrocytes also remove substances such as potassium ions that accumulate in the synaptic extracellular space during neuronal activity. This is important because the prolonged presence of potassium compromises neuronal excitability. To recap, astrocytes maintain an extracellular environment conducive to neuronal activity and protect neurons from the toxic effects of excess glutamate. One might say they are very eco-efficient cells—they practice recycling and, in doing so, produce energy!

Astrocytes have yet another function. We have already described how the energy required by neurons to maintain their excitability, and thus transmit information, is supplied in large part by astrocytes in the form of lactate (neurons also use some glucose, particularly when they are in a resting state). Lactate and glucose are converted into energy in the form of ATP in the neuronal mitochondria (intracellular organelles that serve as ATP-production factories). The problem is that the production of large quantities of ATP by the mitochondria generates highly reactive products—free radicals—which are dangerous to neurons. These reactive molecules must be quickly neutralized. You might say that neurons behave like polluting car engines whose carbon monoxide fumes must be neutralized by a catalytic converter. It turns out that astrocytes produce a molecule called glutathione that acts as a catalyst to neutralize these free radicals (Dringen et al., 2000).

You can see that neurons live a very dangerous life. They produce potentially toxic molecules such as glutamate (when it is present in excess) and free radicals, which can destroy them. Astrocytes protect them from their hazardous lifestyle (Magistretti and Allaman, 2015, 2018; Jourdain et al., 2016).

Neurons Talk to Astrocytes

We've seen that neurons are fed and cleaned by astrocytes and depend on them for their survival. But are astrocytes merely neuronal
housekeepers? Historically it was thought that neurons alone were
responsible for transmitting information throughout the nervous
system, hence the term, "neurotransmission." This neuro-chauvinist
bias was challenged in the early 1980s when researchers identified
various neurotransmitter receptors on astrocytes in vitro, e.g., in
astrocyte cell cultures (Magistretti et al., 1983). These studies suggested that neurotransmission is not the sole province of neurons.
It turns out that neurons can also "converse" with astrocytes! As an
example, let's see what happens when a visual stimulus is transmitted
(Figure 19).

To be certain that neurons communicate with neighboring astrocytes, we must first confirm that the results from in-vitro experiments apply in vivo (in live organisms). A team at the Massachusetts
Institute of Technology studied this in ferrets (Schummers et al.,
2008). The group observed that when they presented an appropriate

Figure 19. When an astrocyte perceives a visual stimulus

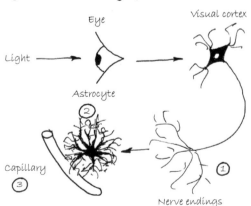

Step 1. Neurons are activated and release neurotransmitters; Step 2. Neurotransmitters
also activate adjacent astrocytes; Step 3. Astrocytes induce changes in blood flow.

visual stimulus to ferrets, both neuronal and astrocytic activity (in the form of calcium waves) peaked in a specific region of this animal's visual cortex.

At first only neurons were activated (Step 1 in Figure 19), but then astrocytes responded a few seconds later, which corresponds to the time required for neurons to release their neurotransmitters (Step 2). As this neuronal response spread to adjacent astrocytes, it caused blood flow changes in the surrounding capillaries (Step 3).

Surprisingly, astrocytes responded very selectively to the information conveyed by neurons, e.g., they responded only to visual stimuli presented with a vertical orientation but not a horizontal orientation, or vice versa.

This experiment, and others like it, demonstrate that an environmental signal first transmitted to a discrete population of neurons is then conveyed to astrocytes. In short,

- neurons transmit information (or orders?) to astrocytes;
- astrocytes are indeed intermediaries between the neuronal command and capillary blood vessels, presumably ensuring that neurons get the nutrients they require to function; and
- individual astrocytes can selectively respond to specific environmental stimuli.

Astrocytes Talk to Neurons

Astrocytes talk to neurons, providing rapid and precise responses—but how do they accomplish this? The conventional notion of neuron-to-neuron communication via synapses has been expanded to include communication between neurons and astrocytes and communication within a network of astrocytes. But the surprises don't stop there; astrocytes release molecules that act on neurons and may even modulate synaptic activity. A new term, gliotransmission (as opposed to neurotransmission), has been coined (Halassa et al., 2007; Araque et al., 2014). Even more astonishing is that glutamate and GABA (gamma-aminobutyric acid)—two neurotransmitters par excellence—are released by astrocytes as well, and, thus, can also be considered gliotransmitters (Parpura et al., 1994; Bezzi et al., 1998; Le Meur et al., 2012).

What takes place between astrocytes and neurons as they converse? It turns out that astrocytes modulate synaptic transmission (Jourdain et al., 2007). In some cases, the glutamate released by a neuron, in addition to acting in a conventional manner on another neuron, interacts with receptors on the astrocytic processes surrounding the synapse (Figure 20). Once the astrocyte is activated by this neuronal glutamate, it releases its own glutamate in turn, which then amplifies the synaptic transmission between neurons. So, astrocytes, through glutamate release, behave like synaptic "turbochargers" (Perea et al., 2009). The astrocyte-released glutamate may also activate and synchronize a group of neurons. A reverse effect is obtained via another pathway: an astrocyte activated by neuronal glutamate may release either GABA or ATP, a molecule which is rapidly transformed into adenosine. Both ATP and GABA exert an inhibitory effect that is particularly potent on the synapses surrounding the first synapse activated. By limiting the excitatory signal to a single synapse, the signal becomes spatially precise (Halassa et al., 2007; De Pitta et al., 2016).

It's evident that the classical notion that information processing relies solely on neuron-to-neuron communication is essentially obsolete. Neuronal and astrocytic networks are engaged in an intense dialogue mediated via chemical signals, i.e., neurotransmitters and gliotransmitters, respectively. In neurons the release of neurotransmitters is triggered by electrical impulses, whereas in astrocytes the release of gliotransmitters is set off by calcium. This neuron–astrocyte dialogue underpins the notion of the tripartite

Figure 20. The tripartite synapse: neurons talk to astrocytes and vice versa

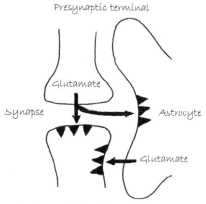

Presynaptic terminal

Glutamate

Synapse

Astrocyte

Glutamate

Postsynaptic dendrite

According to the classical model of neurotransmission, neurons communicate with each other via synapses. A neurotransmitter, glutamate for example, is released from the presynaptic terminal and exerts its effect on the postsynaptic neuron. But according to the tripartite synapse model, a neurotransmitter released by the presynaptic terminal also acts on the astrocyte processes surrounding the synapse. The astrocyte then releases its gliotransmitter, in this case also glutamate, which acts on the postsynaptic neuron. In this latter model of neurotransmission there are three partners: the presynaptic terminal, the postsynaptic neuron, and the astrocyte. Hence the term "tripartite."

synapse, in which the astrocytic processes surrounding the synapse work in concert with the neuron's pre- and post-synaptic elements (Halassa et al., 2007). This three-way dialogue also points to the role of astrocytes as integrators of neuronal activity. Astrocytes can amplify and synchronize the activity of groups of neurons. Conversely, they can also locally inhibit synaptic activity in order to fine-tune signal resolution.

A Conductor of the Neuronal Symphony

The astrocytic network, interconnected by gap junctions, is able to disseminate information about ongoing neuronal activity across a brain region. This signal amplification function sets in motion mechanisms that maintain an optimal environment for neuronal activity. Neurons, it turns out, are also connected in a network, but through their synapses. Returning to the example of Neuron A activating Neuron B (Figure 17), this type of activation recruits hundreds of other neurons connected throughout a network. This in turn elicits calcium mobilization by astrocytes. The ensuing calcium waves trigger a metabolic wave, i.e., astrocytes washed over by this calcium tsunami take up more glucose and "feed" it to neurons in the form of lactate, thus supplying more energy for the activated neuronal network (Figure 21) (Bernardinelli et al., 2004; Charles, 2005). In some instances, however, the astrocytic network will also lift the conductor's baton and synchronize the activity of hundreds of activated neurons. This synchronizing effect helps optimize the processing of information and plays a role in memory mechanisms (see Chapter 3, Figure 25).

Beyond the scope of the cellular mechanisms, which we are starting to grasp, we should really take a look at how the astrocytic network could play a role in higher brain functions. One way to do this in laboratory animals is to block communication across gap junctions, either by using drugs or by inactivating the gap junction's constituent proteins. For example, in transgenic mice the genes

Figure 21. The astrocytic network

GLUTAMATERGIC SYNAPSE   ASTROCYTIC NETWORK   CAPILLARY

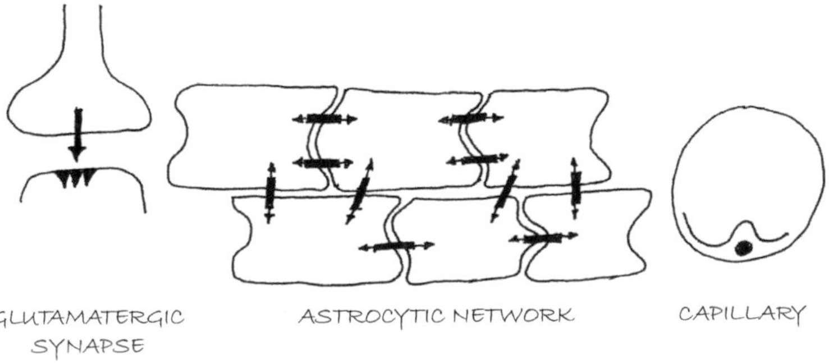

Glutamate released by a presynaptic neuron terminal in the vicinity of an astrocyte can trigger a calcium wave that propagates step by step through the astrocytic network. This calcium wave boosts the initial glutamatergic signal and triggers a metabolic wave that amplifies glucose uptake.

coding for connexins (the proteins that make up gap junctions) were inactivated and the mice displayed abnormalities related to memory and learning (Pannasch et al., 2011). These mice also exhibited motor control disorders.

Regulating Neuronal Plasticity

To survive and thrive, humans must adapt to changes in their environment. This means our brains must be versatile and our neurons able to modify their responses as new information comes in. The processing of new information utilizes existing neuronal circuits and even creates new ones. The capacity of our brains to form an abundance of new and efficient synapses is called neuroplasticity (or brain plasticity).

Astrocytes are involved in the formation of new synapses (synaptogenesis). Synaptogenesis, which is essential to our well-being and survival, takes place

- *in the embryo and the child* while neuronal circuits are developing;
- *at any age* when learning is taking place (see 'Astrocytes and Rat Behavior' p. 44), and
- *any time the brain has been injured* (Clarke and Barres, 2013).

Synapse formation proceeds in three stages:

1. Synapses form in the presence of astrocytes, but remain silent.
2. These silent synapses become mature and functional.
3. The nerve connections, thus established, become stable with the pruning of excess synapses in such a way that only functional synapses remain to form neural circuits.

Even though the mechanisms behind neural circuit formation are not fully understood (formation, maturation, synaptic pruning), it's clear that astrocytes play a preeminent role in the formation and stabilization of synapses in accordance with a pre-set plan (Slezak and Pfrieger, 2003; de Pitta et al., 2016). So, astrocytes, by contributing to the formation of neuronal circuits, certainly participate in developing the overall plan that guides the brain's development. This makes sense given that we know astrocytes influence not only synapse formation (synaptogenesis), but also axon growth (axonogenesis), two components required for the formation of neuronal groups.

A Memory Booster

We've seen that astrocytes help imprint experiences in our brain, as well participate in the physiological mechanisms underlying neuroplasticity, which until recently were considered the sole the domain of neurons. What astrocytic mechanisms are at work here? Returning once again to the simple model of Neuron A talking to Neuron B (Figure 17), we emphasize that the information transmission between these two neurons is not static. Thanks to neuroplasticity mechanisms, the communication between Neurons A and B gains in efficiency as the neurons become more intensely activated (Figure 5). When we have a significant experience to remember, these mechanisms strengthen the connections between thousands of neurons on a lasting basis.

Astrocytes and Rat Behavior

Experiment 1: the more experiences a rodent has, the more astrocytes are found in its brain

Two groups of rats were raised under different conditions: one group in a stimulus-rich environment (with toys, a treadmill, etc.), and the other (the control group) in a stimulus-poor environment (a simple cage). After a while, what was observed in the visual cortices of these animals? Compared to the control group, the rats whose intellect and perceptions were stimulated showed larger and more numerous astrocytes (Sirevaag and Greenough, 1987), as well as an increased number of synapses. This included more synaptic connections with other neurons, an index of increased communication between neurons (Markham and Greenough, 2004; Kleim et al., 2007).

Experiment 2: the more active a rodent is, the more astrocytes are found in its brain

Another group of rats was made to do complex tasks that involved forced or non-forced exercises (for example, using a treadmill) and compared to an inactive control group. The cerebellum of each group was studied—a brain region involved in the control and adjustment of fine movements. The results were similar to those in the first experiment. There was an increase in the number of synaptic connections and in the total volume of astrocytes in the active group. The volume of blood vessels (capillaries) also increased—probably so they could supply additional energy. Thus, it appears that motor learning increases both the number of synapses and the volume of astrocytes (Black et al., 1990; Sirevaag and Greenough, 1991).

In these experiments, astrocytes contributed to neuroplasticity by supporting an increase in the number of synapses and, in doing so, supported enhanced communication within the brain. Thus, astrocyte involvement is critical to the control of synaptic timing, location, number, and plasticity. Through the course of evolution, astrocytes have likely played a key role in developing our tremendous capacity for neuroplasticity and, hence, in developing our impressive set of sensory and motor skills.

Astrocytes release two gliotransmitter molecules (serine and lactate) that ramp up plasticity mechanisms and boost memory. How does this work? These gliotransmitters act on a subclass of neuronal glutamate receptors that go by the unwieldy name NMDA (N-methyl-D-aspartate) receptors (Figure 22). To become activated, this receptor type must bind both the glutamate released by the presynaptic neuron and the serine released by astrocytes in response

Figure 22. The NMDA subclass of glutamate receptors

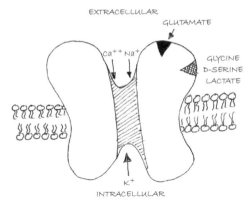

The NMDA subclass of glutamate receptors is composed of two subunits that form a channel permeable to sodium, potassium, and calcium ions. To fully activate this ion channel, both the principal agonist glutamate (a receptor-binding molecule that produces a biological response) and the co-agonist glycine or D-serine, must bind to a selective site on the receptor. The lactate produced by astrocytes amplifies the signal triggered by glutamate and glycine (Yang et al. 2014), which helps to consolidate the changes brought about by synaptic plasticity (Suzuki et al., 2011; Boury-Jamot et al., 2016).

to neuronal glutamate release (Oliet et al., 2006). Once fully activated, the NMDA receptors trigger a reaction cascade within the neuron that renders it more "efficient" on a lasting basis. Like a Cold War power authorizing a nuclear strike, two keys are required to launch the chain reaction underlying memory mechanisms: one is neuronal in origin (through the action of glutamate) and the other is glial in origin (through the action of serine) (Henneberger et al., 2010)

As for lactate, its molecular mechanisms are not as well understood but it also seems to enhance the efficiency of the NMDA receptor (Yang et al., 2014). What is clear is that, if we block the transfer of lactate from astrocytes to neurons, we block the neuroplasticity molecular cascade and prevent the formation of memories (Suzuki et al., 2011). Overall, these experiments demonstrate that lactate, an astrocyte-produced molecule, is required for the consolidation of memory—a far cry from the long-held beliefs that astrocytes serve merely as brain "glue" and that lactate is nothing more than a metabolic by-product (Magistretti and Allaman, 2018).

Neuronal Migration During Development

During embryonic development billions of neurons are generated in a sort of primordial nervous system called the neural tube. Step by step, a rudimentary brain takes shape and neurons migrate to their correct locations—but how do they find their way? In the cerebral cortex, which is the most developed region of the brain in humans, a specific type of glial cell, the radial glia, guides this neuronal migration (Chotard and Salecker, 2004; Merkle et al., 2004; Mori et al., 2005; Clarke and Barres, 2013). The neuronal migration guided by radial glia begins in the area where neurons are produced and continues all the way up to the surface of the cortex. The cerebral cortex is organized into layers in which different types of neurons take up residence. The radial glia form a kind of ladder up which neurons climb to reach the right "floor" in the cerebral cortex (Figure 23). This function of radial glia was proposed more than 130 years ago by two Italian histologists, Camillo Golgi and later Giuseppe Magini, using very basic histological techniques. It wasn't until the development of electron microscopy in the 1970s that their brilliant intuition was confirmed (Bentivoglio and Mazzarello, 1999; Rakic, 2003).

The role of radial glia goes beyond that of serving as neuronal ladders. In fact, once brain development is complete the radial glia divide to produce new cells that can become neurons, astrocytes, or even oligodendrocytes, thereby rounding out the pool of cerebral cortical cell types (Sild et al., 2011). Certain anomalies, often genetic in origin, can affect radial glia and lead to serious brain developmental problems—lissencephaly is one example (Gupta et al., 2002). Anatomically, lissencephaly is characterized by a smooth cerebral cortex in which the usual convolutions (folds) are absent. Clinically,

Figure 23. During development the radial glia serve as ladders for neurons

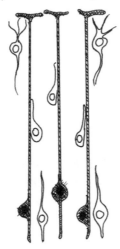

Radial glia (in black) are the ladders up which neurons (in white) "climb" to reach their final destinations.

patients present with severe mental retardation and frequent epileptic seizures. These deficits help us appreciate the vital role of glia in the normal development of our nervous system.

The Formation of New Neurons

Contrary to conventional wisdom, the adult human brain can regenerate neurons. Until the 1990s it was believed that new neurons could not be produced in adults; in other words, once the nervous system has developed, it was thought that neurons lose the ability to multiply. In this scenario, hundred-year-olds would have the same collection of neurons in their brains as they had at age twenty. But the truth is that throughout the brain's developmental stages, i.e., from the time the nervous system assembles up until adulthood, nerve cells continue to proliferate. They are created from stem cells (Appendix VI) that differentiate and transform into various types of neurons (Dimou and Götz, 2014).

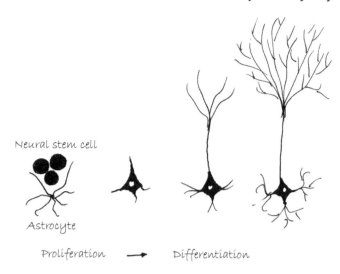

Neural stem cell

Astrocyte

Proliferation ⟶ Differentiation

A little over 20 years ago the scientific community came to accept the notion that stem cells can divide and produce newborn neurons in adult brains (Eriksson et al., 1998; Mori et al., 2005; Aimone et al., 2014). And if newborn neurons exist in human adult brains, why wouldn't they be involved in mental functions?

The question also arises of how to define "true" neurogenesis, in light of recent advances in the field (Aimone et al., 2014; Kempermann et al., 2015)? First, does neurogenesis occur in all brain structures? These days, we think not. As far as we know using biochemical techniques, new neurons are produced in the hippocampus and, to a lesser degree, in disparate brain regions (the olfactory tubercles and the periventricular zone). They may also show up in regions such as the striatum, but this has not been firmly established. A second question is whether the formation of new neurons is physiologically meaningful? The answer seems to be yes. On the one hand, neurogenesis increases during periods of learning and exercise, suggesting that it facilitates human mental functions. It seems that learning and memory storage go hand in hand with the formation of new neurons in the hippocampus. Interestingly, antidepressants also induce neurogenesis and may exert their beneficial effects through this same mechanism (Kato et al., 2013). However, chronic stress negatively impacts neurogenesis (Gould and Tanapat, 1999; Pham et al., 2003; Mirescu et al., 2006; Warner-Schmidt and Duman, 2006; Egeland et al., 2015), as does aging—in very old rats the number of newborn neurons falls by nearly ten-fold.

Astrocytes are involved in the formation of new nerve cells in the adult brain at two levels: (1) astrocytes are generated from neuronal stem cells in the same way as new neurons and (2) they can modulate the differentiation of neuronal stem cells.

Neurogenesis and Memory

Newborn neurons were first described in the 1990s but considered incidental. They were viewed as evolutionary relics, a sort of functionless neuronal appendix. But that's no longer the case. Today there are reasons to believe that newly formed neurons participate in memory processes. For example:

1. It's no coincidence that these new neurons appear in the hippocampus, a structure that plays a vital role in memory production. Ablation of the hippocampus leads to a type of forgetfulness that prevents the memorization of new factual information.

2. Knowing that new neurons are formed, it's reasonable to propose that they are somehow involved in memory processes. Since we know that mature neurons encode the memories of past events, it's logical that new neurons would be involved in the encoding of new experiences. Supporting this hypothesis is the observation that neurogenesis gradually diminishes in aged subjects—and, as we know, memory functions also degrade with age.

Astrocytes in the Service of Neurons?
Neurons in the Service of Astrocytes?

Astrocytes are clearly not neuronal handmaids any more than neurons are astrocytic servants. They work together to provide the flexibility and efficiency required for the nervous system's proper functioning. They make a pair, bound together for better or for worse. But this power couple knows how to work together.

1. *Astrocytes possess the necessary and sufficient properties to ensure the development and survival of neurons.* They communicate speedily with neurons and at a slower pace with other astrocytes over short and long distances.
2. *Neurons and astrocytes converse.* Neurons speak to astrocytes and astrocytes speak to neurons with the same goal: to integrate and synchronize information so that they can ensure the accuracy and specificity of our most complex behaviors.
3. *Astrocytes enable the formation of new synapses during learning. They also play a role in the formation of new neurons in adults* (Sultan et al., 2015). They constitute a reliable and robust network that ensures the integration of the vast array of information conveyed by neurons.

In short, astrocytes and neurons are interdependent. They appear to work synergistically to promote optimal brain functioning. We will demonstrate this principle in Chapter 3 with some select examples.

3
Astrocytes and Behavior

Neurons, Astrocytes, or a "*Pas de Deux*"?

Today, no leading scientist would subscribe to the idea that astrocytes eclipse neurons when it comes to key brain functions—we are still wedded to a neuro-centric hypothesis of behavior. While we may amass arguments to support this hypothesis, the scientific process also obliges us to eliminate the alternative glio-centric hypothesis. For the sake of thoroughness, we must examine arguments that favor a role for glial cells in the shaping of mental abilities, whether they be intellectual or emotional. Once we construct a "glial" theory of brain functioning, we can test—and perhaps demolish—the arguments underpinning it. So then, what is the main scientific evidence supporting the involvement of glial cells in the brain's workings? The goal of this part of the exercise is not to ascribe all mental faculties to glial cells, but rather to lay out the evidence supporting a glio-centric hypothesis, even if that means systematically disproving it later.

So what evidence favors a glial hypothesis? Take any behavior, for example, the act of retrieving the pen from a table discussed in Chapter 1. How does the information flow from the time we first perceive the pen to the moment we grasp it in our hand? The process begins when photons emitted by the pen activate retinal receptors in our eyes. These receptors integrate information related to the pen's shape, color, and location in space. They then activate the neurons of the optic nerve, which in turn relays signals to the posterior region of the cerebral cortex (the visual cortex). Within the visual cortex, electrical signals selectively convey information on the shape, color, and topography of the pen that is then integrated by neurons specific to each of these properties, thereby providing the brain with an accurate image of the pen. But this extraordinary transformation of an object's photon activity into neuronal electrical activity is still not sufficient for us to recognize the pen, let alone ascribe meaning to it. Now, a third class of neurons in the visual cortex comes into play. These latter neurons receive the message that an object (the pen) is perceived, but not yet identified, and nearly instantaneously they receive signals from other brain regions holding stored memories related to features such as form, color, and texture. With this additional information, the brain is able to recognize that the object is indeed a pen.

Up to this point we have summarized, in broad strokes, the process of perceiving and recognizing the pen. Assuming that the pen is recognized as such, it will now be used for writing. To write, one must rapidly perform a series of movements, both simultaneous and consecutive, that involve numerous muscles. In fact, there is a motor control program in the anterior region of the brain (the frontal cortex) that immediately gets to work coordinating the hand-grasp movements required to pick up the pen. This involves not only the fingers, but also the entire upper limb, not to mention complex adjustments in body posture. The various neuronal circuits coordinating these complex and precise movements are quite well understood.

Although the neuronal components underlying the visualization (perception) and utilization (action) of the pen are largely understood, there is less certainty as to what occurs between perception and action. Why does this object function as a pen and not as a pencil or crayon? The implication is that the perceived object (the pen) has a function. What is it to be used for? How does it differ from other pens? These are unresolved lines of scientific inquiry. As we see, the propagation of a visual message within the brain, from the time of perception to the triggering of motor behavior, involves a much more complex neuronal operation than just the routing of afferent (input) and efferent (output) information. Some people consider this processing of information within the brain to be an intractable mystery, while others see an opportunity to apply a heuristic model. Remember that neurons and glial cells intervene at each stage in the routing of the pen-object information—the glial cells include the oligo-dendrocytes that form the myelin sheath responsible for acceler-ating neuronal conduction; as well as the astrocytes, which are in contact with hundreds or even thousands of neurons, along with their tens of millions of neuronal connections.

Whatever interpretation put forward, whatever model proposed, we still lack conclusive evidence as to how our brain, stocked as it is with neurons, generates a mental picture of a pen that enters our consciousness. We have not yet considered the swarm of astrocytes that interact with the neurons bearing the image of the pen. There are three possible scenarios to consider regarding the role of astro-cytes in generating the behaviors that prompt us to pick up the pen:

1. *Neurons do all the work, including the conveying and interpreting of information.* According to this scenario neurons "think" and astrocytes play some ancillary role, but are clearly servile to neurons.

2. *Neurons provide the flow of information, but are unable to pro-duce a thought*, i.e., the mental picture and manner of using a pen. We could attribute this thought-producing function to other players, e.g., glial cells. In this scenario, the pen's image and the action of using the pen are conveyed by neurons, but it's left to astrocytes to recognize the pen, understand its useful-ness, name it, enjoy it, keep track of it, judge it against previous experiences, direct it in a personalized way to write on paper, and make sense of the text that it produces.

3. *Neither neurons nor astrocytes perform these tasks on their own, but instead work in concert.* Neurons provide the sensory input and the motor output, but only the neuro-glial pair is able to integrate, synchronize, interpret, recognize, and act. In a nut-shell, to think!

Astrocytes and the Integration of Neuronal Messages

There are two reasons to believe that astrocytes are involved in how neurons receive and produce information:

1. *The first—anatomical—is that the contiguous arrangement of astrocytes allows them to form a three-dimensional continuous cellular network* made up of interlinking communication compartments that appear to be complex enough to integrate a wealth of information.

2. *The second—physiological—is that astrocytes generate signals that establish bridges between different neuronal networks*, which then become functionally interconnected. How is this done? The astrocytes interact with several thousand nerve connections belonging to neighboring neurons. They receive a copy of the information being transmitted from neuron to neuron through the myriad of neuronal synapses. We know the mechanism involved: intracellular calcium is converted into calcium oscillations in the form of calcium waves that are synchronized throughout the entire astrocytic syncytium via gap junctions. If the calcium signal is weak, only nearby synapses are modified; but if the signal is strong, distant synapses are also affected (Newman et al., 2003).

This is how the electrical and chemical information originating in neurons is integrated within the astrocytic syncytium and synchronized thanks to the calcium oscillations produced by gap junctions (Volterra and Meldolesi, 2005) (Figure 24). Unlike the "neuronal doctrine" model, this arrangement confers the physiological advantage of neuronal mechanisms that are integrated and synchronized in time and space—which, though it may not be enough to support a mental process, is certainly an essential requirement.

How this integration, via electrochemical synchronization, allows the material substance of the brain to generate something as ethereal

Figure 24. Astrocytes are integrators and synchronizers

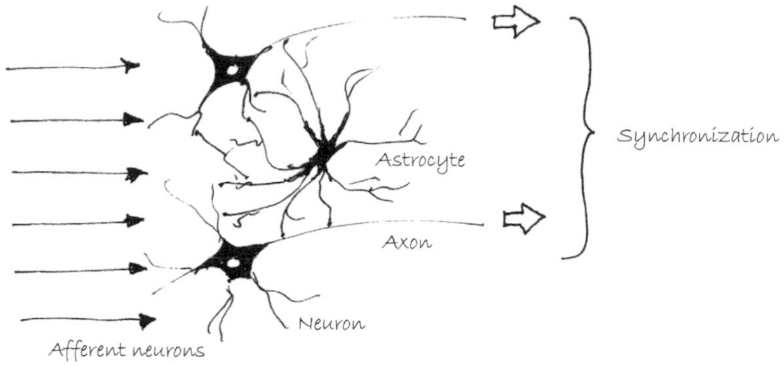

as thought is still a mystery to ponder. Nonetheless, if there is a type of cell that should take part in this alchemy, it is the astrocyte. Why do we believe this is so? What arguments favor a role for astrocytes in the genesis of behaviors?

Learning and Memory

Learning depends on a number of factors, such that if any are missing learning cannot occur. These factors include one's memory capacity, as well as the four "pillars" of brain functioning (Agid, 2013). The proper functioning of these pillars (alertness, attention, motivation, mood) is indispensable. If I am asleep, distracted, apathetic, or depressed, my ability to retain information and learn is compromised. This is all instrumental when it comes to utilizing our catalogue of past experiences, i.e., our inventory of recent and older memories, whether we are conscious of them or not. When we have new experiences, they combine in complex ways and integrate into our existing memories. But, how does this newly learned information integrate spatially into neuronal populations spread across the brain, i.e., into specific regions throughout the cerebral cortex? It's still a mystery. Temporally, the integration of information takes place over hours, days, months, or years—way beyond the milliseconds it takes to transmit a neuronal signal.

To succeed at learning one must memorize; but how? Although memory is still the most studied function of the brain, we are still uncertain about its underlying mechanisms. Given that electric signal transmission is the basis of all brain activity, it's natural to assume that signal-carrying neurons are on the front line when it comes to memory processing.

Remarkably skilled at ensuring the transmission of electrical signals, neurons are essential to the process of conveying the information to be memorized to the appropriate brain areas—areas specialized in neuronal plasticity (see Figure 5), i.e., that have the capacity to repeatedly and preferentially channel newly learned information through the same neuronal circuits and reconfigure nerve connections in the course of repetitive learning. The credo these days is that neurons "learn" via the mechanisms of synaptic plasticity. This presupposes that a huge number of nerve impulses will, by an astounding temporal and spatial coincidence, converge on the same neuron at the same time so that there is a perfect synchronization of neuronal activity waves that are also in phase so as to enable the flawless integration of messages contained in the information arriving from different areas of the brain. Hard to imagine, but then why not?

Attributing the memory phenomenon solely to neurons raises questions. How does the unbridled flow of information acquired over time imprint into neuronal groups and create a durable memory trace? How can neuronal point-to-point transmission explain the brain's ability to integrate the large amount of information pouring in from diverse regions? How do we account for the memory process—which functions both over the short and long terms (from a few hours to several months)—by just considering neurons, which operate within a very brief timeframe ranging from a few milliseconds to a second? Is there a mechanism that ensures the integration of messages with respect to both time and space? In fact, there is. There are glial cells!

Barring the possibility of a heretofore undiscovered molecular mechanism within neurons (which may still be discovered down the road), astrocytes are the best candidates to manage, or at least help to manage, a complex cognitive process like memory. Besides their numbers and surprising biochemical properties, astrocytes possess the necessary anatomical and physiological properties required for memory handling, which involves broad spatial integration and regulation over a long period of time.

From an Anatomical Perspective

Human astrocytes are unique in their level of sophistication when compared to the astrocytes of other animals, including rodents. They are much larger (20 times bigger in humans than in mice— see Figure 8); they come into contact with an impressive array of neighboring synapses (a single human astrocyte contacts about two million synapses, as compared to only 100,000 in rodents), which explains their ability to integrate a considerable amount of afferent (incoming) information; and finally, they put out rich and complex connections that strengthen their natural capacity to integrate signals (Bushong et al., 2002).

All this qualifies astrocytes as potential mnemonic assistants. According to the prevailing theory, memory formation depends on neuronal plasticity mechanisms that optimize the transfer of information between neurons at the synaptic level in a sustainable manner (called long-term potentiation or LTP; see Figure 5). But as discussed above, neurons may not be the only ones involved in forming memories. An experiment conducted by Christian Henneberger and his team in 2010 points to a supporting role for astrocytes in this process. In their classic experiment they elicited LTP in the hippocampus through repeated electrical stimulation of presynaptic neurons. The passage of the electrical signal continuously through the same route produced "synaptic strengthening"—the so-called LTP. The researchers modified the

activity of a single astrocyte with a micropipette and observed the
following:

- High-frequency stimulation of Neuron 1 (the presynaptic
  neuron) provoked elevated calcium levels in the astrocyte,
  which in turn enhanced synaptic transmission and brought
  about LTP in Neuron 2 (the postsynaptic neuron).
- If they inhibited the increase in astrocytic calcium levels with an
  appropriate solution, LTP was abolished (Figure 25).

Assuming that LTP in neurons is a true indicator of memory processing, this work demonstrates that memory production can progress, or not, depending on what astrocytes "decide." This suggests
that even though neurons play a role in memory, they must do so in
concert with astrocytes.

Figure 25. Astrocytes as mnemonic assistants

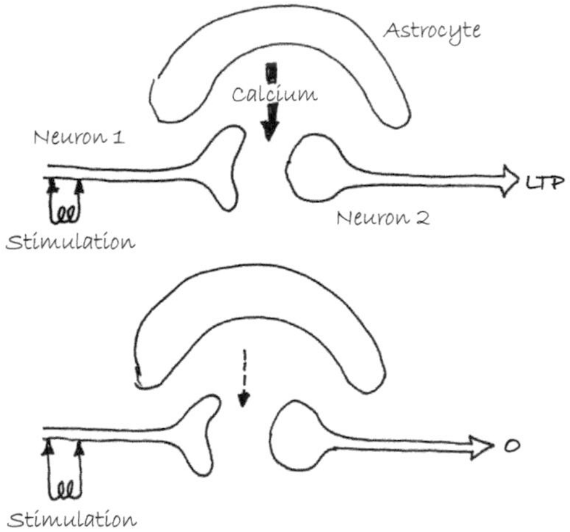

From a Physiological Perspective

Astrocytes do not generate electrical signals like neurons, but they do possess capabilities for information integration that are missing in neurons. Astrocytes can

- regulate synaptic transmission between neurons;
- modulate synaptic strength;
- pair up various synaptic domains into functional assemblies; and
- promote the creation of new neurons.

These properties of astrocytes (Figures 17–25) allow them to work together to open up more avenues for information storage and boost memory storage capacity. It's hard to imagine how neurons could join forces to provide equivalent functions.

Many recent experiments indicate that astrocytes participate in the learning process, and particularly in memory encoding and consolidation. Cell physiology experiments confirm that astrocytes may either activate or abolish LTP (Figure 25) in animals. For example, when the brains of neonatal mice received human astrocyte grafts, the mice were able to learn faster and they demonstrated an increased capacity for memory compared to the control group, which received only mice astrocyte grafts (Han et al., 2013) (Figure 26).

We draw two main conclusions from this type of experiment:

1. Given that human astrocyte grafts boost the memory capacity of mice, we can assume that these improvements derive not only from the presence of neural networks, as sophisticated they are, but also from the presence of the transplanted astrocytes. In other words, the new astrocytes played a role in memory formation.
2. The fact that transplanting human astrocytes, rather than mouse astrocytes, led to improved memory function, suggests

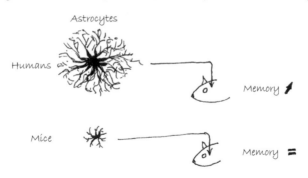

Figure 26. Human glial cell grafts improve rodent memory

that human astrocytes have properties specific to humans, properties not inherent to mouse astrocytes. Why are human astrocytes better memory boosters than mouse astrocytes? Is it because human memory is superior to mouse memory?

At any rate, these results are consistent with what we know about the properties of astrocytes in humans and in mice. The human astrocytes are larger and have more extensive ramifications (Figure 26) than do the astrocytes of their mouse counterparts. This is further evidence that astrocytes have taken on a greater role in neural processing with evolution (Han et al., 2013).

Could Consciousness be Glial?

There are as many different definitions of consciousness, a human being's highest mental function, as there are experts in the field. Does consciousness involve access to a mental picture? One's ability to interact with an ever-changing environment? Is it one's ability to relate to oneself? Or is it thinking about a thought, a perception, or an action concerning oneself? Whatever it is, a conscious brain can process an array of related afferent information in a unified way, even though the information may appear to be incomprehensibly fragmented. This mass of information must be integrated somewhere in the brain and assembled into a coherent mental corpus. However, although the anterior portion of the brain, the frontal cortex, plays a key role in managing consciousness, it is probably generated by other anatomical entities scattered throughout brain that link together their information, a phenomenon called "binding."

Which of the brain's cellular components bind all this information? Would it be the neurons? Perhaps, but consider that—given what we know about the anatomical and physical-chemical properties of neurons, and despite the complexity of neuronal afferents and efferents—the fundamental behavior of neurons, i.e., neurotransmission, is quite stereotyped. Aside from neurons, there is a sea of astrocytes that receive a copy of the sensory information. They are certainly excellent candidates, since each one contacts several tens to hundreds of thousands of nerve endings, and each is able to produce slow calcium waves that could ensure the integration and synchronization of all neuronal information. Without the presence of the astrocytic syncytium and its capacity for integration, how else could the specificity of a signal be preserved amid this neuronal clutter? Despite remarkable progress in modern psychology, notably due to brain imaging technology, it bears repeating that we still don't have answers to basic questions like: how do we make sense of our world and ourselves; or what are the physiological foundations of the "self"? Nonetheless, the presence of astrocytes is likely a necessary, if not sufficient, condition for explaining the existence of consciousness.

Love and Glial Cells

Who could seriously claim to understand the basics of love or attempt to define it precisely? Well, Stendhal perhaps, but even then . . . Hence the interest in coming up with simple models that account for the rudimentary mechanisms underlying this complex yet universal behavior that varies so much from one individual to the next. Where might we seek out such models? The fly provides one, of course! In one experiment, the sexual behavior of flies was studied by artificially modifying the activity of their brain astrocytes (via a selective mutation within astrocyte cells) without affecting the neurons (Grosjean et al., 2008). The researchers noted that the genetically altered fly, which is normally heterosexual, no longer distinguished between male and female sexual partners. Without going so far as to say the fly became bisexual due to an astrocyte mutation, one can state that the drastic genetic modification of its glial cell population sparked a major behavioral change.

Glial Cells and Romance

When a male fruit fly (*Drosophila melanogaster*) courts a female fly, his behavior is pre-programmed. As described by the authors (Grosjean et al., 2008), "to confirm his suspicions and to test whether she is sexually receptive, he will tap her with his foreleg (to evaluate nonvolatile pheromones via chemoreceptors on his leg), sing a species-specific courtship song (by extending and vibrating a wing), and lick her genitalia (to sample pheromones). If she is acceptable and does not reject him . . . he will mount her, curl his abdomen and attempt copulation."

In this experiment the researchers produced a transgenic (genetically altered) fly with a mutation affecting an amino-acid transporter exclusive to astrocytes, and with no neuronal involvement. The researchers observed that this altered transporter triggered male copulation behavior toward females and males alike. It's as if the males, while still heterosexual, behaved bisexually. Furthermore, when the mutation was homozygote (an individual carried two identical copies of the mutant gene), the male carriers of the mutation more eagerly seduced other males than did the heterozygotes (carriers of one mutant copy of the gene in question); who, nonetheless continued to be more attentive to other males than to females.

In other words, the male fly carrying the mutation is no longer able to discriminate between male and female. Through a series of additional experiments, the authors isolated several chemical factors that could mediate this behavior, including one involved in taste and smell called 7-tricosene.

They concluded that a specific modification of an astrocytic function can change a behavior as sophisticated as sexual choice. But to suggest that astrocytes play a role in romantic love and the choice of a sexual partner in humans would require a leap of faith best avoided at this point in time.

Glial Cells Got Rhythm—A Circadian One!

All animals, including humans, experience constant change in their environment; such as changes in light intensity, temperature, and the presence or absence of predators. Our behavior in any particular situation is influenced by the time of the day, whether we are awake or asleep, or if it is light or dark. These behavioral adaptations that occur in any 24-hour period depend on a day–night cycle called the circadian rhythm. The circadian rhythm obeys an internal clock that provides most animals with an innate sense of time and makes it possible for them to constantly adapt to changes over time. It is vital, for example, in determining our periods of activity and rest.

We are beginning to understand more about the neural circuits that generate this circadian pacemaker. They involve small areas such as the pineal gland or the hypothalamus that are found at the base of the brain. We are also identifying the neurotransmitters involved in this neuronal network. Until recently no one had proposed a role for astrocytes in the regulation of circadian rhythms, but we now understand that the modification of certain molecules within astrocytes can change these rhythms. For example, genetic techniques can be used to manipulate whether or not astrocytes express a selective enzyme (*Ebony*) that disturbs the fruit fly's circadian rhythm (Emery and Freeman, 2007).

Such studies demonstrate that astrocytes physiologically modulate the neuronal circuitry directly involved in circadian-dependent behaviors. This type of work holds out promise for those suffering from circadian rhythm disorders (nighttime insomnia, daytime sleepiness, etc.) who don't respond to conventional treatments. It's reasonable to speculate that drugs acting on astrocytes may one day help restore normal sleep patterns.

Starry Nights: Sleep and Glial Cells

We spend about a third of our lives asleep. Though we may not be conscious; sleep is, in fact, a very structured state of mind. Electroencephalography (EEG), which measures the brain's electrical activity, reveals two phases of sleep: slow wave sleep, during which the cortical neurons synchronize their activity and produce large low-frequency waves; and paradoxical sleep (Jouvet, 1967, 1992), so called because brain activity resembles the waking state, even though the subject is deep asleep! Our dreams take place during this latter phase. Although we still do not understand the reason for sleep, it is undoubtedly important to the memorization process and the metabolic recovery of the brain.

Given the involvement of astrocytes in synaptic transmission, it's no wonder that more and more studies point to a role for astrocytes in regulating the sleep–wake cycle. Recent analyses using electron microscopy have shown that astrocytic processes surrounding synapses retract during slow wave sleep. This diminution in the astrocytes' synaptic coverage has functional implications. One of the main functions of astrocytes is to "pump out" glutamate released into synapses (see Figure 18), which has the effect of limiting this neurotransmitter's spatial reach. If there is a pulling back in synaptic coverage by astrocytic processes, glutamate "overflows" the synapse and can activate many more synapses. This leads to a synchronization of the activity of numerous neurons—precisely what we observe during slow wave sleep. During wakefulness we see the reverse effect; the synaptic coverage by astrocyte processes becomes more pronounced, which locally limits the action of glutamate (Tononi and Cirelli, 2014;

Bellesi et al., 2015). More than 100 years ago, Santiago Ramón y Cajal intuited a dynamic role for astrocytic processes in sleep, but he erred in the details. He thought that during sleep, astrocytic processes invade the synapse to block transmission, acting like circuit breakers between neurons (Fellin et al., 2012; Tso and Herzog, 2015). In reality, during wakefulness, astrocytic processes surround synapses and optimize the postsynaptic action of glutamate; whereas during sleep, astrocytic processes retract and allow glutamate to spill out of the synaptic zone, which diminishes the effectiveness of signal transmission—in other words, the opposite of what Cajal had in mind (Figure 27).

Figure 27. Astrocytes free up synapses and allow sleep

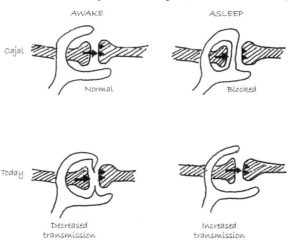

It appears that sleep is an active process; it relies, in part, on the elevated expression of genes specific to astrocytes during the sleep cycle, especially genes linked to energy metabolism and the transfer of lactate to neurons. This lactate transfer is required for memory consolidation, a known function of sleep (Petit et al., 2013).

The above examples demonstrate the strong physiological link between neurons and astrocytes; one cannot function without the other. Weighing this evidence, we now consider the question of whether neurons are able to generate our behaviors on their own, or if the neuron–astrocyte partnership is the essential unit governing human activities.

4

Astrocytes and Neuropsychiatric Disorders

The Complexity and Specificity of Brain Diseases

The disciplines dealing with brain diseases are essentially "neuron" oriented, as their names indicate. For example, we say *neuro*-logy, *neuro*-anatomy, *neuro*-physiology, etc. A more inclusive nomenclature would take into consideration the entire family of *nervo*us system cells, including glial cells. It would be more accurate to employ terms like *nervo*-logy, *nervo*-anatomy, *nervo*-physiology, etc. since we know several brain diseases are truly *nervo*-logical, in the sense that they arise from glial cell dysfunctions.

To determine whether glial pathology can cause neuronal dysfunction, let's apply the same reasoning we would use to demonstrate that neuronal pathology can lead to glial cell dysfunction. If we accept the proposition that all brain functions, from the most basic to the most complex, involve a neuronal–glial partnership, then it follows that both cell types should be involved in pathologies of the nervous system. We know that astrocytes account for more than half of all human brain cells, so we would naturally expect any brain affliction at least partially to impact the workings of astrocytes. An important question to address is, for any particular malady: does the primary disruption occur in neurons, or in glial cells—or in both? And in this third scenario, which cell type comes first in the cause-and-effect sequence: neurons or glial cells?

For purposes of simplicity, let's draw a distinction between neurological and psychiatric disorders. According to this somewhat

artificial distinction, neurological diseases (so-called organic disorders) are treated by neurologists and neurosurgeons, whereas psychiatric disorders include the "disturbances of the mind" handled by
psychologists and psychiatrists. The division of these diseases into
two different categories will prove less useful over time as our understanding of their underlying mechanisms evolves.

Neurological Disorders

These fall into two categories:

1. *Those associated with other organs of the body* (tumors, infec-
   tions, hemorrhages, infarctions, trauma, poisoning, etc.) that
   are still prevalent in developed countries, but pose particular
   challenges in poor or developing countries.
2. *Those associated only with nervous system tissue* (neurodegenerative diseases, multiple sclerosis, epilepsy, brain malforma-
   tions, migraines, etc.) and that constitute a true socio-economic
   tragedy in any country owing to their elevated incidence in in-
   fants (often with debilitating and sometimes fatal consequences)
   and in the elderly (increasingly prevalent due to increased life
   expectancy).

It is striking that, with few exceptions, multiple sclerosis for one,
we always think in terms of neuronal dysfunction. The presence of
astrocytes in the vicinity of a brain tissue injury is most often viewed
as part of a compensatory response, a reactive gliosis, i.e. a nonspe-
cific change of glial cells in response to a central nervous system
injury—but the reality is more complicated.

Psychiatric Disorders

Psychiatric disorders account for nearly 30% of the annual health-
care budget in Europe (Morris et al., 2016). Psychiatry is medicine's

most challenging discipline. The array of symptoms is more complex than in other disease groups and the mechanisms that give rise to them are poorly understood.

- *To clinicians*, these disorders reflect subtle emotionally driven mental functions that are multifaceted, insofar as they manifest as a diverse set of symptoms that change over time. Managing psychiatric disorders is a challenge!
- *To scientists*, the brains of afflicted psychiatric patients may look "normal." There are no "holes" in the nervous tissue, such as one sees after a hemorrhagic stroke brought on by the rupture of a cerebral artery; there are no lumps due to tumors; and there is no loss of neurons, as is the case in an Alzheimer- or Parkinson-type neurodegenerative disease. The underlying causes of psychiatric afflictions are not visible to the naked eye, rather they occur at the molecular level. Psychiatric research presents a challenge!
- *To the layperson*, psychiatric disorders remain mysterious. Apart from those who believe the "sick" spirit resides outside the physical body (dualism), the origin of psychiatric diseases such as depression, psychosis, and obsessive–compulsive disorder are unclear. Perhaps environmental factors (education, family life, work, living conditions) predominate—a reasonable assumption in mild cases of anxiety or depression. Or perhaps genetic predispositions (through family transmission) are the chief culprits, as is the case with Huntington's disease or bipolar depression. Science cannot always discriminate between the respective contributions of our genes and the environment (nature versus nurture) to these disorders—this, despite the dazzling progress in research in this area. Pinning down the causes of psychiatric disorders remains a challenge!
- *Finally, there is the realm of politics.* Unfortunately, tension persists between those espousing clinical treatments based on purely psychological approaches (psychoanalysts) versus those embracing biologically based treatments that draw on the latest advances in the clinical and experimental neurosciences. The

discord continues, despite more open-mindedness on both sides of the divide. This lack of a meeting of minds has had negative implications for research, where one finds a great many high-quality clinical studies (e.g., the testing of new drugs) but a dearth of lab-based experimental studies. It's a challenge to find consistency among the various trends in psychiatry!

Is there evidence pointing to an unequivocal role for astrocytes in the genesis of psychiatric disorders? There certainly is! The best known astrocyte-related disorder is Alexander disease. Alexander disease is a rare genetic disease that from the time of birth results in an enlarged head size (megalencephaly), epileptic seizures, and profound intellectual disabilities. This horrendous affliction is exclusively astrocytic in that it can be traced to a mutation in the gene that encodes GFAP (glial fibrillary acidic protein), which is expressed in astrocytes (Sofroniew and Vinters, 2010; Messing et al., 2012). The abnormal version of this protein accumulates in astrocytes, which then hypertrophy—neurons are not involved in this process. Death follows on the heels of profound suffering.

If we needed proof that glial dysfunction can sow behavioral chaos, then Alexander disease fits the bill. In other situations, neuronal distress is accompanied by a proliferation of astrocytes, or abnormalities in their functioning, or both.

To better evaluate the contribution of glial cells, and more specifically astrocytes, in nervous system pathology, it's necessary to distinguish what is causal from what is secondary in the pathological process—a correlation doesn't necessarily imply a cause-and-effect relationship.

One cannot determine the cause of a pathological process by simply observing sick patients or analyzing diseased nervous tissue obtained after death or via biopsy. Hence, the importance of experimental studies that selectively modify or interrupt astrocyte activity without modifying neurons. This was how researchers demonstrated that a genetic mutation produced exclusively in astrocytes leads to the distress or death of nearby neurons (Figure 28).

Figure 28. When astrocytes go silent, neurons die

Altered astrocytes can contribute to neuronal dysfunction and elicit a characteristic clinical picture. Cell culture experiments have shown this connection for three well-described disorders: (a) trisomy 21, which is caused by a third chromosome at position 21 instead of the usual pair and gives rise to the intellectual disability known as Down Syndrome; (b) a mutation on the X chromosome that causes Rett Syndrome, observed in young girls and associated with intellectual deficits and respiratory problems; and (c) Fragile X Syndrome, which involves a gene mutation that causes synaptic dysfunction and consequent intellectual deficits (Clarke and Barres, 2013).

Interestingly, if we place astrocytes afflicted with any one of these three mutations in an appropriate culture medium along with healthy neurons, the latter do not develop normally, i.e., the development of their nerve endings is stunted.

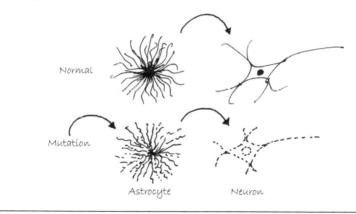

Given the above observations, which point to a close interaction between neurons and astrocytes, it's fair to assume that the functioning of neurons is intimately influenced by nearby astrocytes and, as a consequence, dysfunction in one cell type will impact the other. Thus, any nervous system disorder labeled as "a *neuro*-nal disease" would, of course, entail changes in the functioning of astrocytes as well. Should we go so far as to speak of *astroglial* diseases? Not necessarily. If we carefully review what is known about the underlying mechanisms of nervous system disorders, we see that the role of astrocytes can be

- a positive one, e.g., as protectors or helpers, as in Parkinson's disease; or
- a negative one, e.g. their direct involvement in Alexander disease or their indirect role in certain dementias; or, as is usually the case,
- equivocal, or even inconclusive.

These results suggest that an abnormality in the development of astrocytes can play a role in the occurrence of neurodevelopmental disorders. Maybe we should consider targeting astrocytes, along with neurons, in the quest to cure these nervous system ailments.

Neurodegenerative Diseases and Glial Cells

Unfortunately, there is no universally recognized conceptual framework to turn to when characterizing the involvement of astrocytes in the onset of nervous system disorders, hence the arbitrary one proposed below. These diseases appear with varying frequency, depending on whether they preferentially affect the cerebral cortex, basal ganglia, cerebellum, spinal cord, or peripheral nerves.

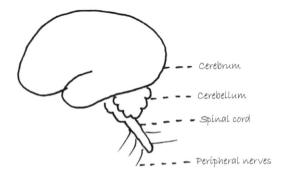

The Cerebral Cortex Under Siege

Alzheimer's Dementia: Do Astrocytes Protect or Harm?

As we age, most if not all of us will complain of memory loss at some point. This is a common clinical picture and one that can elicit fear, but in the vast majority of cases it's nothing more than "benign forgetfulness"—we fail to recall a word or name during a conversation. It's annoying but, voilà, later that evening the evasive word pops back into our consciousness, maybe while eating dinner or in the midst of a totally unrelated exchange. This is a classic "memory lapse" experienced by folks who are distracted, overworked, or depressed. Though it may arouse concern, it is, in fact, inconsequential. However, some people really do lose their memory permanently. In these pathological cases, patients do not generally complain about memory loss. Curiously, they are unaware of their oversights, a condition called anosognosia. Yet over time their memories inevitably fade, first the most recent ones and eventually the older ones too.

We might think this is commonplace in very aged persons. Isn't it normal for the elderly to undergo memory deficits? It would seem to go hand-in-hand with their slow pace of walking and the feeling of being less sure on their feet as their sense of balance erodes. What, in fact, is the difference between "normal" memory loss and pathological memory loss in the elderly?

From a clinical perspective, in the former case, an individual can recall an event when reminded (their memory can be jogged); but in the latter case, the memory is definitively lost—this is often indicative of Alzheimer's disease, a diagnosis that must be confirmed with specific tests.

Alzheimer's disease is the most prevalent degenerative dementia, affecting about 5.5 million people in the USA and nearly 800,000 in France. Furthermore, this scourge appears to be worsening. It is a condition whose incidence is on the rise as life expectancy increases. It is a chronic disease that causes patients to lose autonomy and become dependent on caretakers, with all the attendant socio-economic consequences. In spite of its seriousness, we still have no therapeutic treatment to halt the progression of this "disease of the century"! Consider this fact: more than 1,000 new clinical trials have been conducted worldwide over the past ten years, with a failure rate of 99.6%!

What changes do we observe in the brain of an individual with Alzheimer's disease versus a normal brain? A 100-year-old in good mental health loses few if any neurons, even though the number of nerve endings diminishes progressively with age, like a tree losing its leaves. As a consequence, communication within the brain becomes increasingly difficult (Figure 29). By contrast, in the unfortunate Alzheimer patient we see the inexorable loss of the neurons themselves.

This neurodegenerative disease is distinctly different from the normally aging brain in that

- *the loss of neurons, while slow, is faster than in the course of normal aging,* and

Figure 29. The neurons in the brain of an adult, an elderly person, and an Alzheimer's patient

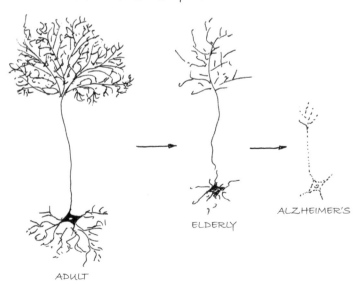

ADULT

ELDERLY

ALZHEIMER'S

- *the loss of neurons is selective* (it is not a global loss of neurons). It first impacts a specific population of neurons in a specific brain region (the hippocampal region of the temporal lobe, in the case of Alzheimer's disease).

What happens to the astrocytes surrounding the degenerating neurons in Alzheimer brains? Oddly, there is no clear-cut answer to this question because researchers studying this type of pathology have focused primarily on the fate of the deteriorating neurons and are less interested in what befalls astrocytes (Fu et al., 2014). In this affliction neuronal loss is accompanied by the accumulation of a characteristic protein fragment (beta-amyloid) inside senile plaques. These protein bodies are easily observed under the microscope (Figure 30).

Figure 30. Alzheimer's disease: the overwhelmed astrocytes

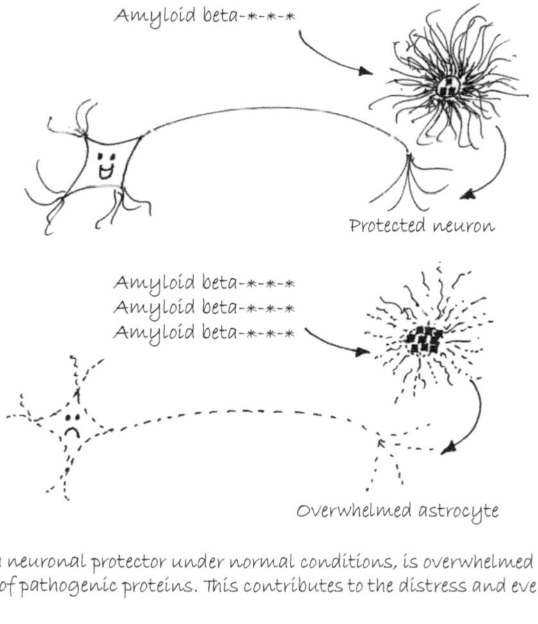

Amyloid beta-*-*-*

Protected neuron

Amyloid beta-*-*-*
Amyloid beta-*-*-*
Amyloid beta-*-*-*

Overwhelmed astrocyte

The astrocyte, a neuronal protector under normal conditions, is overwhelmed by the accumulation of pathogenic proteins. This contributes to the distress and eventual demise of neurons.

Two facts have emerged regarding this disease. First, astrocytes specifically bind, internalize, and degrade beta-amyloid protein (Allaman et al., 2010). Everything proceeds as if the astrocytes were exerting a protective role. But the situation is more complex because, by dint of "swallowing" this pathological protein, they lose their ability to provide the metabolic support neurons require and, as a result, the energy-deprived neurons begin to degenerate (Allaman et al., 2010). Second, the involvement of astrocytes in this pathological process is even more direct. Only astrocytes produce the ApoE (apolipoprotein E) molecule, which plays a key role in triggering Alzheimer's disease. ApoE, a lipid that exists in several biochemical forms, is a major indicator of one's predisposition to developing Alzheimer's disease (Figure 31). Individuals who carry the ApoE4 variant of ApoE are eight times more at risk for developing this disease than those carrying other variants (Liu et al., 2013). The fact that all ApoE variants are produced by astrocytes indicates the importance of these cells in the genesis of this awful disease.

Figure 31. How a lipid helps trigger Alzheimer's disease

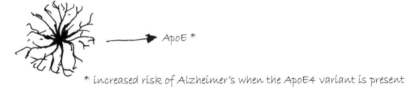

ApoE *

* increased risk of Alzheimer's when the ApoE4 variant is present

Frontotemporal Dementia

There are other forms of degenerative dementia besides Alzheimer's disease; the most common being frontotemporal dementia (FTD), in which lesions predominate in the frontal and temporal lobes. This disorder, previously known as Pick's disease, is often hereditary—which is seldom the case for Alzheimer's disease—and is also characterized by progressive memory impairment. Yet, this isn't the most troubling aspect of this disease. FTD patients are overwhelmed by behavioral disorders that rapidly become unbearable, i.e., mental agitation, lack of social inhibitions, inappropriate gestures, etc. What are the mechanisms involved?

Under normal conditions, astrocytes produce very small amounts of a class of proteins called tau. In this disease there is a specific mutation within chromosome 17 that triggers an inappropriate accumulation of abnormal tau proteins within astrocytes (Forman et al., 2005). We are uncertain as to whether the loss of neurons in this disease is due to neuronal dysfunction per se, or to dysfunction in the neighboring cells—astrocytes, for example.

Animal studies have been of great value in working out the details of this disease. In transgenic mice, the selective introduction into astrocytes of a mutation identified with FTD leads to the overexpression of tau proteins only in astrocytes, i.e., not in neurons. However, this inappropriate accumulation of tau proteins in astrocytes eventually leads to a loss of neurons, concomitant with a decrease in the cognitive abilities of the mutated mice (Forman et al., 2005). The dementia observed here is a result of neuronal dysfunction, but the underlying cause is astrocytic! Despite the fact that the molecular mechanisms of neuronal degeneration are poorly understood, this type of experiment demonstrates that pathological astrocytes can elicit a neuronal disease (Figure 32).

Figure 32. Frontotemporal dementia

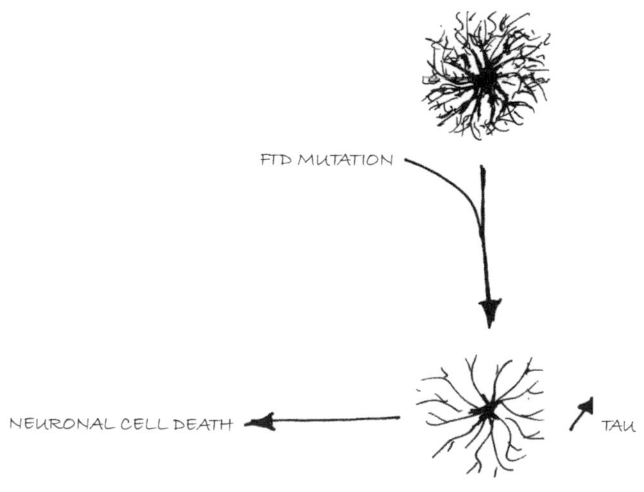

FTD MUTATION

NEURONAL CELL DEATH

TAU

A mutation in astrocytes leads to neuronal cell death

Basal Ganglia Involvement in Parkinson's Disease: Astrocytes as Helpers and Protectors

Parkinson's disease (we should say "diseases" because this term includes a number of heterogeneous disorders grouped under the rubric "parkinsonian syndrome") is a neurodegenerative disease that involves a progressive and selective loss of dopaminergic neurons (neurons utilizing the neurotransmitter dopamine) deep beneath the cerebral cortex. It turns out that this small population of neurons (about 600,000 in number or 0.001% of all neurons in the brain) act as a sort of on/off switch for our behaviors. As these neurons degenerate, patients struggle to initiate movements they previously performed without conscious effort. French neurologist Jean-Martin Charcot remarked that Parkinson's patients are "condemned to a life of voluntary movements."

There is as yet no cure for this disease, nor do we understand its causes, except for some rare hereditary forms. We do understand quite a bit, however, about the biochemical mechanisms underlying the degeneration of dopaminergic neurons. Yet questions remain: Why are dopaminergic neurons selectively destroyed, and no other types of neurons? And why, among the larger population of dopaminergic neurons, do only some degenerate?

Astrocytes may provide some answers to the above questions. For example,

1. the varying density of astrocytes under normal conditions may explain the vulnerability of dopaminergic neurons, and
2. the accumulation of astrocytes in the vicinity of damaged dopaminergic neurons may signal a protective role for them, in which they ward off neuronal degeneration, as suggested by the neurologist Philippe Damier and his team (Damier et al., 1993) based on their observations of autopsy tissue (Figure 33).

Figure 33. Parkinson's disease: astrocytes as protectors?

1. *In normal individuals* there is a low density of glial cells in the brain region most susceptible to Parkinson's disease, i.e., the dopaminergic neurons of the lateral part of the substantia nigra, a basal ganglia structure of the midbrain. In contrast, glial cell density is six-fold greater in the medial part of the substantia nigra, an area where dopaminergic neurons are spared the ravages of Parkinson's disease. These results suggest that, under normal conditions, the glial cells present are able to protect dopaminergic neurons from degenerating.

2. *In the Parkinson's patient* the opposite is true: in the lateral part of the substantia nigra, the more neurons degenerate, the more glial cells accumulate; conversely, in the medial portion of the substantia nigra, where neurons are more resistant to degeneration, fewer glial cells accumulate. It seems that, in the disease state, glial cells proliferate mainly in regions where dopaminergic neurons are in trouble, as if to protect them. We know, moreover, about the lifesaving power of glial cells in regions where there is neuronal cell death, i.e., they can release trophic factors that ensure the survival of neurons.

This research on post-mortem brains suggests, even if it does not prove, that in the non-diseased state, glial cells are able to protect neurons and that in the diseased state they do their best to limit neuronal cell loss.

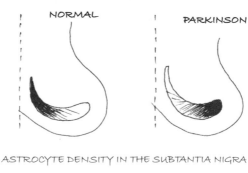

ASTROCYTE DENSITY IN THE SUBTANTIA NIGRA

☐ LIGHT   ■ HEAVY

The Spinal Cord and Amyotrophic Lateral Sclerosis

Amyotrophic lateral sclerosis (ALS, also known as Lou Gehrig's disease) is a disease that causes paralysis, along with rapid and progressive muscle wasting; death may occur within months of onset. ALS is a neurologist's nightmare, since no tools exist to halt its progression. It occurs sporadically and is hereditary in 10% of the cases. It is characterized by the loss of all neurons involved in motor control, including the major efferent nerve pathway that descends from the motor cortex to the spinal cord, the pyramidal tract, and the motor neurons that convey signals from the spinal cord to the muscles. The hereditary forms of ALS arise from a number of mutations; one of particular interest involves the gene that expresses SOD1 (superoxide dismutase 1), an enzyme that produces an especially toxic reactive oxygen species (Cleveland and Rothstein, 2001).

Here we have an archetypal "neuronal" condition and yet its underlying cause is associated, in part, with dysfunctional astrocytes. Astrocytes possess a glutamate carrier that rapidly eliminates this neurotransmitter from the synapse once its excitatory role is complete (see Figure 18). In doing so, this uptake mechanism protects neurons from the toxic effects of excess glutamate. When the carrier is missing, the glutamate reuptake mechanism is dysfunctional and glutamate surges to toxic levels (Danbolt, 2001). In ALS there is a selective loss of this specifically glial glutamate carrier in the motor cortex and spinal cord, precisely in those areas where the first and second contingent of motor neurons—the ones lost in this disease—originate (Seifert et al., 2006; Philips and Rothstein, 2014; Radford et al., 2015).

Two other observations point to a key role for glial cells in ALS. First, researchers observed that the level of a protein that enables astrocytes and oligodendroglia (the glial cells that form the myelin sheath) to supply lactate to neurons falls considerably in ALS patients, which suggests a role for this protein in ALS pathogenesis (Lee et al., 2012). Second, in studies where astrocytes obtained from deceased ALS patients were placed in contact with normal neurons, the latter degenerated. This suggests that the pathological astrocytes released toxic substances that damaged neurons (Ferraiuolo et al., 2011).

As a consequence, neuronal degeneration in this hereditary form of ALS is due to

- the overproduction of certain oxygen species, associated with a mutation in SOD1;
- a deficit in the ability to remove glutamate from the synapse, leading to excito-toxic levels of glutamate;
- a decreased availability of lactate due to astrocytic malfunctioning; and
- the production of toxic compounds by astrocytes.

Is this disease neuronal or astrocytic in origin? A study using mice carrying the SOD1 mutation addressed this question. This mutation can be expressed in neurons, astrocytes, microglia, or all three. It seems that, individually, these mutations, whether neuronal, astrocytic, or microglial will not precipitate neuronal cell death. The latter is observed only when the SOD1 mutation is ubiquitous, meaning it is expressed in all three cell types at once (Lobsiger and Cleveland, 2007).

Yet again, this experiment demonstrates that a neurodegenerative disease, once viewed as strictly neuronal, can also be considered astrocytic. Neuronal and astrocytic mutations work together to unleash this disease, whose evolution is further accelerated by the presence of a microglial mutation.

These types of observations should motivate medical scientists to also target astrocytes in their quest to mitigate the damage wrought by this horrendous disease!

Is Spinocerebellar Ataxia Neuronal, Astrocytic, or Both?

Spinocerebellar ataxia (SCA) is a group of neurodegenerative diseases originating in the cerebellum that cause, to varying degrees, a progressive deterioration in motor coordination. The symptoms usually manifest as abnormal movements, sensitivity disorders, muscular weakness, and impaired vision. This clinical picture is genetic in origin and at least 25 mutations are involved.

One of these ataxia syndromes, SCA7, is fortunately quite rare. It is a so-called dominant transmission (carriers of the gene have a 50% chance of passing it on to their offspring) brought about by a mutant sequence in the gene that encodes the ataxin-7 protein. The mutant gene produces ataxin-7 proteins that have excessive repeats of the amino acid glutamine (polyglutamine tracts). Why mention this variant of ataxia? In individuals whose astrocytes express this mutation, once cerebellar neurons begin to die, balance difficulties associated with this disease ensue. But, in fact, the mutation may be expressed in both astrocytes and neurons. So which mutation is harmful, the one in astrocytes or the one in neurons? It seems the disease is most serious when the mutation is expressed in astrocytes (Figure 34)!

Figure 34. Cerebellar ataxia: an astrocytic or neuronal disease?

The mouse provides an experimental model for evaluating these types of mutations. A harmful mutation is introduced into the ataxin-7 gene. But thanks to some molecular "cut and paste," the mutation can be selectively introduced into cerebellar neurons (Purkinje neurons) or into neighboring astrocytes (Bergmann glia), or into both (Lobsiger and Cleveland, 2007).

In all three cases, the mutation causes cerebellar Purkinje neurons to degenerate. However, the greatest effect is observed when the mutation is introduced into the Bergmann glia. The impact is minimal when the mutation is introduced only into Purkinje neurons. A mutation introduced into both cell types yields a result that falls between the two extremes.

Once again, here we have a disease universally considered to be neuronal in origin, since there is a massive loss of neurons. Nonetheless, it can be triggered by astrocytic dysfunction, and, in fact, it is more lethal when the mutation occurs in astrocytes rather than in neurons. This surely points to the sway the "lowly" glial cells hold over the "lofty" neurons.

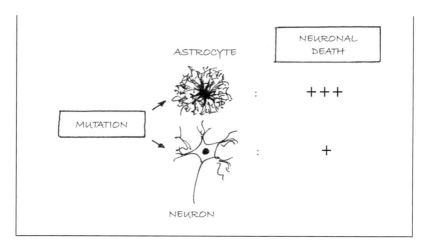

Encephalopathies

In a number of acute ailments, astrocytes swell up with the entry of extracellular fluid, which can trigger some dire scenarios—the best-known example is cerebral edema. This form of edema accompanies severe head trauma and can turn fatal as the brain swells within the rigid structure of the skull, which of course does not expand to accommodate it. Edema may also follow a stroke brought on by the occlusion of a cerebral artery, an event that often has debilitating consequences, as in cases of hepatic encephalopathy (see below). Either way, astrocytes that were once protective quickly lose their ability to help out.

Hepatic Encephalopathy in Alcoholics

Hepatic encephalopathy, a condition seen in advanced-stage alco-holics with liver cirrhosis or in some patients with viral hepatitis, leads to an impairment of consciousness and possible coma. The coma is the result of cerebral edema that brings about an increase in intracranial pressure (Figure 35).

Figure 35. Hepatic encephalopathy

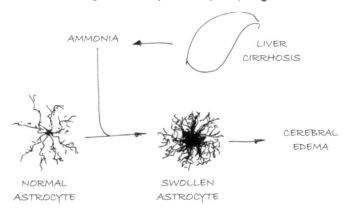

Astrocytes are central to this affliction because brain edema is, for the most part, caused by swollen astrocytes (Lizardi-Cervera et al., 2003). Why is it astrocytes that swell up and cause edema, but not neurons? Astrocytes distend in response to the production of unusual quantities of ammonia (NH_3) by the sick liver. In their protective role, the job of astrocytes is to clear out excess ammonia, but they are soon overwhelmed by this herculean task and become harmful themselves.

The astrocytes take up ammonia and convert it into the amino acid glutamine by combining it with glutamate. The accumulation of glutamine in astrocytes causes them to swell, which in turn leads to a generalized edema that is toxic to all brain cells. In other words, in this pathology, astrocytes that normally act as neuronal protectors by absorbing toxic compounds turn lethal once they have surpassed their detoxification capability.

HIV Encephalopathy (AIDS–Dementia Complex)

There is every reason to believe that under certain conditions, when they are overwhelmed, astrocytes can kill neurons. This is the case with human immunodeficiency virus (HIV) encephalopathy, also known as acquired immune deficiency syndrome (AIDS)–dementia complex (Churchill et al., 2009). In the days when AIDS patients had no access to effective treatments, about 20% developed rapidly progressive dementia. Microglial cells, not neurons, are the prime targets of the HIV virus.

Oddly, we do not know how infected microglia cause dementia, given that the neurons appear unscathed! So, we can only speculate that

1. microglia become activated by inflammation and secrete potentially toxic molecules (the cytokine TNF-alpha, for example); and that
2. these microglia then activate astrocytes, which lose their capacity to recycle glutamate. This leads to the excessive accumulation of this neurotransmitter. The high levels of glutamate fuel excitotoxic effects that alter neuronal functioning. Thus, infected microglial cells, by triggering a series of complex neurotoxic cascades that in turn disrupt astrocytes, are the linchpin in the evolution of HIV dementia (Figure 36).

Figure 36. HIV encephalopathy

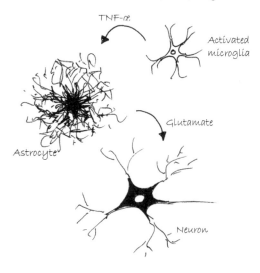

A Lethal Assault on the Nervous System

Stroke

Stroke is the third leading cause of death in humans and the leading cause of serious long-term disability. When we say "stroke," we must distinguish between the relatively rare brain hemorrhage (from the Greek *haimorrhagia*, meaning "bleeding violently") caused by the rupture of an artery and the more frequent cases of ischemia (from the Greek word *iskhaimos*, meaning "stopped blood") that result from a blocked artery. Acute ischemic stroke is brought about either by the gradual clogging of an artery, which eventually becomes completely blocked (this is the case with atherosclerotic plaques) or by the sudden occlusion of an artery by a blood clot (an embolism) that has migrated to the brain from the heart (something that can happen in cardiac arrhythmia patients).

This total blockage of an artery abruptly cuts off the energy supply to the nervous tissue supplied by that blood vessel, with devastating effects. The brain, more than any other organ, is a major energy consumer, and so a sudden interruption of vital nutrients (especially glucose and oxygen) leads to the asphyxiation of nervous tissue. Within just a few minutes, an area of necrotic tissue develops around the blocked vessel. Then the death of nerve cells spreads outwards with decreasing intensity into the remainder of the infarcted area (the penumbra) (see Figure 37).

Today, the best treatment for ischemic stroke is a rapid one that prevents the damage from spiraling out of control. When a stroke occurs, it may be possible to dissolve the clot, a process called thrombolysis or thrombolytic therapy, provided the patient is seen within the first three hours. Beyond that timeframe, no effective neuroprotective treatment is available.

Over the past 20 years more than 1000 therapeutic trials have been conducted but, aside from thrombolytic therapy, none has yielded concrete benefits in the treatment of stroke (Barreto et al., 2011). And yet, the underlying mechanisms of ischemic stroke are well

Figure 37. An ischemic stroke (cerebral infarction)

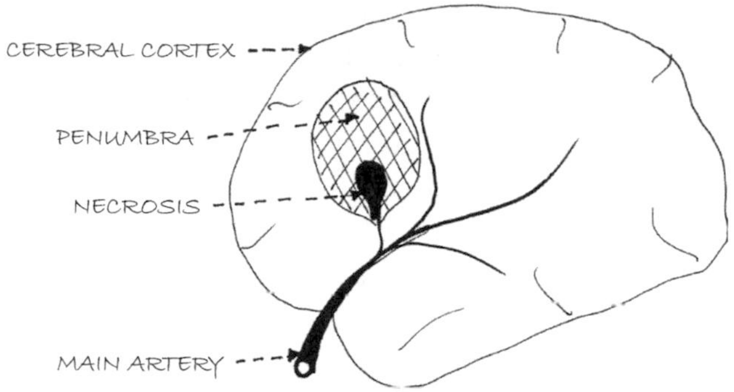

understood. Why this dismal state of affairs? It is because the pathophysiology of brain ischemia is a complex affair that does not progress in a straightforward manner with regard to

- *its evolution in time*; meaning it depends on whether the victim is examined in the first few hours (acute phase), in the first few days (subacute phase), or in the first few weeks (chronic phase) following the stroke, and
- *the severity and size of the infarction*, from its necrotic center (where the nerve cells have died) out to the periphery (the penumbra, where the nerve cells are affected to varying degrees).

In any case, brain ischemia leads to a great deal of cellular and molecular turmoil (Zhao and Rempe, 2010). At the cellular level, neurons are first in the line of fire, while at the same time astrocytes swell, blood vessels alternate between vasoconstriction and vasodilation, the blood–brain barrier is breached, and the whole region is engulfed in edema (Kimelberg, 2005). At the biochemical level, astrocytes are activated but their role remains ambiguous. On the one hand they can cause harm (the secretion of free radicals and cytokines, and they participate in an inflammatory reaction) but on the other hand they can be protective (the secretion of trophic factors). Often, they play both roles at the same time: for example, astrocytes can release glutamate, which is toxic at high concentrations (harmful), while they can also scoop up excess glutamate (protective).

Going forward, how will we identify a therapeutic strategy for ischemic stroke that mitigates its tragic sequelae, i.e., disability and even death? Current neuroprotective thrombolytic treatments, effective in the earliest hours of the stroke, are focused on protecting neurons. If our goal is to discover a therapeutic agent that reduces the size of brain infarctions, it would seem logical to target all of the nervous tissue harmed by the ischemia, i.e., not only neurons, but also glial cells, and especially the astrocytes. Realistically speaking, therapeutic strategies that do not minimize astrocyte death are unlikely to be effective. One need not be clairvoyant to predict that the best therapeutic strategies for the future will be the ones that preserve the function and ensure the survival of the entire "neuro-gliovascular" unit, i.e., neurons, astrocytes, and capillaries.

Spinal Cord Injury

In industrialized countries, the prevalence of spinal cord injuries is about 500 per million inhabitants (900 in the USA) and these are usually related to road accidents (Singh et al., 2014).

A spinal cord cut at the level of the mid-back will result in paralysis of the lower limbs (paraplegia), whereas a cut at the level of the neck paralyzes all four limbs (quadriplegia). When the spinal cord is completely severed, no recovery is possible, but if the cut is partial, some degree of recovery may be possible.

The consequences of a cut are tragic because the spinal cord—normally the width of a large pencil—now has a gap that separates millions of axons (descending from brain neurons to muscles) from their extensions and it is virtually impossible for the severed axons to bridge the gap and reconnect with themselves.

There is also the problem posed by scar formation. The wound-healing process, which in our bodies is beneficial to almost any injury, can actually be harmful to the nervous system. This is the case for spinal cord injuries, where the scar is glial in nature—a gliosis is a form of healing specific to injured nervous tissue (Silver and Miller, 2004). Astrocytes at the injured site convert from a non-reactive to reactive state and form a kind of fibrous barrier (Sofroniew and Vinters, 2010) (Figure 38).

In principle gliosis is beneficial, in that it allows for the encapsulation of an infected site, but more often than not it is an impediment because the fibrous tangle of astrocytes prevents the damaged neurons from regrowing. In these latter cases, the presence of astrocytes is deleterious, in fact very much so! Not only do excess astrocytes prevent neuronal regrowth, but these over-reactive astrocytes can also secrete toxic substances that kill neurons (Lobsiger and Cleveland, 2007). Recently, however, some hope has appeared on the horizon for spinal cord injury patients. Researchers have demonstrated that some degree of regrowth across the glial scar is possible in the presence of various neurotrophic (growth) factors; including, paradoxically, some glia-derived factors (Anderson et al., 2018).

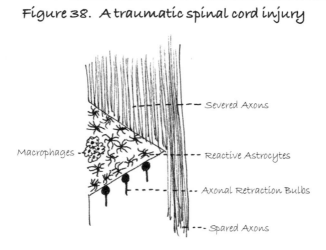

Figure 38. A traumatic spinal cord injury

What is true for spinal cord injuries is also true for any nervous system insult (trauma, infarction, infection, etc.) that causes the destruction of neuronal axons. Whereas activated astrocytes help limit the extent of the damage, they can also be harmful when present in excess (gliosis scar formation) or when malfunctioning. This is a therapeutic issue to be aware of when considering the repair of injured neurons or new strategies for managing brain and spinal cord trauma.

Epilepsy and Astrocytes

Epilepsy is a public health crisis that affects about 50 million people worldwide (Duncan et al., 2006; World Health Organization, 2019). The disease varies in severity from a momentary lapse in consciousness to grand mal seizures that trigger a prolonged loss of consciousness accompanied by convulsions. Epilepsy has dramatic consequences for those affected, impacting their personal, social, and professional lives. Although remarkable therapeutic progress has been made in recent years, 30% of all epilepsies still defy medical treatment.

This intriguing affliction is caused by the abrupt discharge of a group of neurons that behave erratically, like a sort of short circuit. The neuronal storm that ensues may remain focused at its point of origin or, more often than not, may spread throughout the entire cortex and precipitate a loss of consciousness. The triggering factor that gives rise to this abnormal neuronal discharge is generally thought to be the excessive "synchronization" of a group of neurons, but the exact cause, apart from some select cases such as a brain tumor, is not always identifiable.

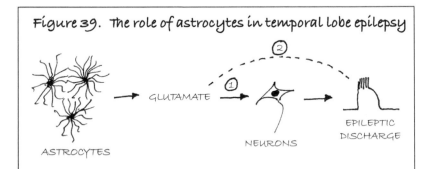

Figure 39. The role of astrocytes in temporal lobe epilepsy

A textbook example of temporal lobe epilepsy occurs in individuals who suffer from hippocampal sclerosis, a pathology that manifests as clusters of altered astrocytes that form a gliotic scar, a process known as gliosis. How can we find out what these astrocyte clusters are doing? The classic experimental model for epilepsy involves using various techniques to set off epileptic discharges in hippocampal tissue slices in culture (pathway 1). What is surprising is that an epileptic discharge, thus triggered, will continue in the absence of neuronal activity (pathway 2). The discharge activity can be blocked by tetrodotoxin, a sodium channel blocker neurotoxin found in pufferfish that is widely used in this type of research (Tian et al., 2005). An in-depth analysis of the mechanisms at play here reveals that epileptic discharges arise in the surrounding astrocytes. Even more interesting is that this astrocyte-based epileptic activity can be suppressed by anti-epileptic drugs that act directly on astrocytes. This study, as well as several others, demonstrates that some forms of epilepsy can be attributed to dysfunctional astrocytes, which means that these glial cells are potential targets for new anti-epileptic treatments.

It is striking, nonetheless, that the spark igniting an epileptic seizure in the cerebral cortex, although neuronal, is accompanied by massive gliosis, i.e., a dense accumulation of more or less normal glial cells. In particular, this is the case for certain focal epilepsies, i.e. those restricted to a specific a brain region, generally in the temporal lobe. Although we understand little about gliosis, we can assume it plays some role in initiating and spreading epileptic activity within nervous tissue (Figure 39).

Another example showcases the role of astrocytes in epilepsy and at the same time provides an explanation for a particular dietary approach to treating this disease, i.e., the ketogenic diet. Since the 1920s we've known that some types of drug-resistant epilepsy can be controlled by replacing all forms of dietary carbohydrates (sugar, bread, starch) with fats (butter, cream, oils), especially in children. On a diet highly enriched in fats, the body produces ketones (byproducts of the breakdown of fatty acids) that enter the brain and replace the lactate that astrocytes normally produce from glucose. Somehow these ketones become fuel for neurons and render them less excitable (Sada et al., 2015).

Psychiatry's Challenging Battle Against Depression

Are astrocytes involved in mental afflictions as complex as depression? And if so, how to go about demonstrating this? Is it better to study depressed individuals or to rely on animal models of depression? Actually, both approaches have proved fruitful (Elsayed and Magistretti, 2015).

There are two complementary approaches to studying the connection between astrocytes and depression in humans. The first involves studies on the post-mortem brains of depressed humans.

Some noteworthy observations have been made using this approach:

- There is a reduction in astrocyte density in the amygdala (a brain region involved in strong emotions such as fear) in the brains of depressed subjects (Sanacora and Bansr, 2013).
- A reduction in astrocyte-related markers is a key feature of a major depressive disorder. In a 2013 review article, Rajkowska and Stockmeier cited decreases in five astrocyte-associated markers, among others: (1) GFAP mRNA—a protein specific to astrocytes; (2) certain gap junction proteins; (3) aquaporin water channels; (4) transporters for excitatory amino acids such as glutamate; and (5) various enzymes (glutamine synthetase, for example). But there is an inherent problem: the results are scattered and unrelated to one another and, thus, hard to interpret.

A second approach to studying depression is to administer strong antidepressants to depressed patients and look for any changes in glial cell metabolism. It is known that the administration of effective doses of antidepressants promotes the formation of new neurons in the hippocampus from primitive glial cells (Santarelli et al., 2003), so it's conceivable that deficient neurogenesis plays a role in depression. But the connection between depression and the malfunctioning of astrocytes, or any other glial cell type, is not so straightforward. In "depressed" mice, researchers are able to assess various astrocytic markers. For example, in a mouse model of depression (stress induced after a social defeat) concentrations of ATP (selectively secreted by astrocytes) were found to be low and the administration of ATP relieved this form of mouse depression (Cao et al., 2013). Interestingly, a more recent animal model study demonstrated that lactate—which, as we have seen in Chapter 2, is mainly produced by astrocytes—can exert antidepressant-like effects (Carrard et al., 2018).

A few words of caution here: these studies still require confirmation because there are numerous biases inherent in human studies and animal models of depression. In human studies, the sampling conditions during autopsy are variable and can distort the data; the diagnosis of depression is not based on uniform criteria and may differ from one individual to the next; the diagnosis is usually retrospective and approximate; and although the brains come from suicidal subjects, not all suicidal victims were necessarily depressed, even if that is generally the case. In animal studies there are multiple models of depression, all of which are approximate (how is one to know if a mouse is truly depressed?), and they are only remotely related to what we customarily call depression in humans. In all cases, due to the variability and disparity of the results, we need to be cautious when we interpret them.

Astrocytes: New Targets for Drug Therapy

Astrocytes are integral partners to neurons in maintaining a healthy nervous system, so we have every reason to expect that effective therapeutic treatments for nervous system disorders should also take astrocytes and their functions into account. Accordingly, we encourage researchers not to overlook these cells in their quest for new therapeutic drugs.

If we reflect on the failure of so many clinical trials to identify any effective treatment for stroke and neurodegenerative diseases, it's evident that researchers are missing something. We have clearly shown that astrocytes maintain intimate ties with neurons, so when the former are damaged or destroyed in a pathological process, the effect is devastating to their neuronal neighbors. So why would researchers only target neurons in their efforts to save neurons? Why not prioritize saving astrocytes as well? The fact is, we don't always know the initial trigger for neuronal loss—is it a neuronal event or an astrocytic one? We firmly believe that it's time to ramp up research on those astrocyte-mediated pathways that could be at the origin of certain neuropathological conditions (Barres, 2008; Finsterwald et al., 2015). The connections between the astrocytic network and overall brain functioning and consciousness have already been well demonstrated by the actions of several drugs that disrupt this network. For example, take widely used general anesthetics such as propofol, halothane, and isoflurane. They mediate their effects by inhibiting the gap junctions that serve as the astrocytes' intercellular communication conduits (Mantz et al., 1993; Wentlandt et al., 2006; Liu et al., 2016).

It is our hope that, in the future, our colleagues in the pharmaceutical industry will develop drugs that not only target neurons, but astrocytes as well (Figure 40).

Figure 40. Will future drugs for nervous system diseases target neurons? astrocytes? Or both!

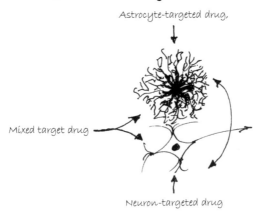

Astrocyte-targeted drug,

Mixed target drug

Neuron-targeted drug

The studies cited above point to the undeniable physiological reality that astrocytes are indispensable partners to neurons and contribute, one way or another, directly or indirectly, to various nervous system pathologies. As we shall see, the ramifications of the myriad interactions between these two cell types are profound, especially as concerns our efforts to understand the underlying mechanisms of nervous system disorders and their treatments.

5

Towards a Revolution in Neurobiology and the Treatment of Neuropsychiatric Disorders

How Can We Discuss the Brain's Role in Behavior If We Ignore Half of It?

Having taken a lengthy journey through the brain's workings, it's time we returned to the question at hand: how to talk about the basis of thought without considering half the brain's cells, i.e. the glial cells? This question has loomed in the background for well over a century, since the time glial cells were first discovered in the mid-1800s. Why only in the past two decades have scientists begun to ponder the role that glial cells perform in shaping our mental functions?

It's as if our concepts about brain functions, from the smallest (molecular) to the largest (socio-behavioral) scale, were hopelessly mired in a neuro-chauvinistic bog. We have suffered from a tunnel vision wholly focused on neurons and have remained blind to the multidimensional nature of brain functioning, which requires its full spectrum of cell types. Why for so many years have scientists side-lined the idea of a physiological role for glial cells in the creation and control of behavior? There are two reasons, one based in science, and the other "psychological," in the broadest sense of the word.

First, as mentioned earlier, the main reason glial cells, especially astrocytes, had escaped scientific notice for so long is because they produce no detectable electrical activity. Hence, they were considered non-excitable and researchers believed that they could not communicate with one another. Today we know better. Glial cells communicate with each other and with their neighboring neurons extensively—clearly they are up to something! In fact, glial cells are useful and even critical to the proper functioning of the brain. They are intricately woven into a vast neuronal–glial communication network. The novelty of this concept means that scientists still consider it radical to assert that the physiology of behavior does not depend solely on neurons—nor solely on glial cells for that matter! In fact, it is the synergy inherent in the neuron–glial pair that holds the key to unlocking the enigma of how thought is created.

The second reason is also scientific, in a manner of speaking, but it does not concern scientific knowledge per se, but rather the attitude of scientists. We scientists have an unfortunate tendency to base our hypotheses on results that bolster our own firmly held beliefs. "A conviction is a disease," noted the artist Francis Picabia. Scientific conclusions are based on facts but despite this we are sometimes influenced by doctrines not wholly grounded in scientific findings. Still, there is no reason to engage in sterile polemics by insisting that two scientific concepts, the one neuronal and the other glial, are mutually exclusive—let's keep an open mind! We know researchers tend to push their favored hypothesis and downplay ones they view as incompatible. That's how the system works, alas. Still, it's a pity that to successfully surf the prevailing ocean of scientific dogma, one must dodge iconoclastic waves.

We wonder—while poking a bit of fun at our profession—what most researchers perceive when they look back over their scientific output? Perhaps they recall creative ideas, accompanied by a great deal of trivia and a few ludicrous moments. They will have had, for sure, some key scientific insights; but also many laborious, not always rewarding, routines to perform. They may have experienced powerful streams of consciousness, maybe not always noteworthy, but perhaps a few that gave rise to trendy ideas that glistened in the light and were seized upon by the scientific media. Only rarely does one experience a volcano of creativity that spews out a torrent of unexpected discoveries. More often than not, the scientific process proceeds by a succession of incremental steps, one following the next, always treading the same conceptual path. The system compels it!

One could say that neuroscience operates as an ideology, perpetuating itself through professional organizations, doctrines, and its disciples. Usually well-intentioned, occasionally with a tinge of arrogance; scientific stakeholders reproduce, with striking tenacity,

similar abstracts, with similar content, with similar bibliographies—
supplemented, whenever possible, by a few private communications.
By dint of repetition, this body of work risks becoming gospel. We
know, since we ourselves have contributed to it.

Even while in the grip of mainstream thinking, a non-conformist
urge can suddenly seize one's mind. We have persistent doubts, for
sure; suppose what we take for gospel is wrong? But how does one
decide to deviate from established dogma? How does one summon
the force to break old habits and routines? It would seem to require
a courageous or enlightened individual. This is a scenario we have
often pondered and what prompted us to write this book.

The questions future researchers face could well be very different.
In much the same way that Claude Bernard, in his day, revolution-
ized notions about the physiology of thought with his concept of the
milieu intérieur, it would behoove us to think less in terms of neural
regions with their nerve connections and intercellular communications, and more in terms of molecular-based functions. In line with
this latter scenario, learning—along with memorization and all consequent thoughts and activities—will be viewed in the context of
their fundamental physicochemical properties (Figure 41).

Figure 41. At what level should we study mental phenomena? At every level!

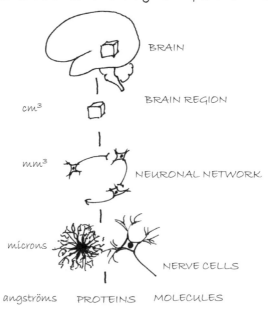

Are Neurons the Leading Contenders for Explaining Behaviors?

Neurons, On Their Own or Collectively, Lack What We Would Call a "Behavior Function"

It's best to avoid over-arching statements like "serotonergic neurons (those utilizing serotonin) are involved in sleep" or "dopaminergic neurons (those utilizing dopamine) are the pleasure neurons, etc." It is like asserting that the Avenue de l'Opéra in Paris is the Opera because it leads to the opera house. An avenue is just a thoroughfare with no dance- or music-related functions. Likewise, let's not assume that neurons or groups of neurons produce thought by virtue of their ability to convey impulses.

Put differently, is the brain's most celebrated cell type, the neuron, the best candidate for producing a phenomenon as intangible as "thought"? It would count as an extraordinary metamorphosis for the neuron, a biological material, to "secrete" something that is colorless, odorless, flavorless, and invisible, but not necessarily inaudible—insofar as when we quiet our minds we can hear the thoughts of others.

Nature does provide us with examples of such metamorphoses. For example, energy in the form of heat is transformed by steam engines into movement and electromagnetic waves are converted by X-ray machines into images. The main difference between the brain and these types of machines is that, in the case of the machines, we understand the finished product (sound, electricity, movement, etc.), whereas we have no firm grasp of what thought is composed of. For now, the material nature of thought remains unidentified—presuming, of course, that intellectual activity and emotions are indeed substance-based.

How Do Communication Pathways (the Neurons) Produce Thought?

Our brain starts by perceiving. In the visual system, for example, the retina perceives the image of an object, which it then transforms into electrical currents. These currents in the form of impulses are conveyed via a succession of neural circuits to the visual cortex in the back of the brain. Neurons are also involved at the other end of this process, since once the brain has processed the perceived information, the decision to act requires the involvement of other neural circuits. This decision-to-act message, formulated in the frontal cortex, is conveyed via the major motor pathway to the muscles, which then carry out the desired behavior.

The enigma here is a difficult but not intractable one. It concerns how the brain processes information, from the time a visual stimulus is perceived at the back of the brain to the generation of a motor command in the front of the brain. Today, scientists seeking to solve this mystery rely on experimental data and theoretical models. The idea that neurons create thought derives from what we know about their capacity for memorization. This is the so-called Hebbian theory, which postulates that neuronal groups are able to learn—hence the quip "cells that fire together wire together" (see Figure 5). If neural circuits store information, inventory it, and enable its recall, we could say that they have indeed memorized information—this in itself amounts to a form of thinking. So, if the Hebbian theory of learning is correct—which we believe it is—it leaves us with the impression that we are closer to explaining the nature of thought. But we must distinguish between content and support. There is no disputing the fact that neurons support the thought-creation process. It is also likely that they transport the content of various mental representations. What is less certain is whether or not they actually transform physical phenomena into ideas and concepts. For the uninitiated this may sound strange; after all, if a neuron "is," therefore it "thinks"—but perhaps not!

If we assume that mental phenomena are strictly neuronal in origin, does it follow that a rudimentary thought is like a reflex, albeit a complex multistep one—different, say, from the simple knee-jerk reflex? If this were the case, we would expect a perceived stimulus to automatically elicit a stereotypic motor response. But how would that explain abstract thoughts, theoretical reasoning, value judgments, or other mind creations? Therein lies the rub!

Drawing on numerous experimental observations, scientists have proposed theoretical models, some of which have been partially validated. Without getting into the details, these theories are all neuron-oriented, whether they concern reentry neurons (Edelman, 1992), proposals related to the access to consciousness (Llinas, 1991), or the conscious neuronal workspace (Appendix V). As sophisticated as these models are, they remain flawed. Their experimental confirmations are imprecise, they deal with vast regions of the cerebral cortex rather than well-identified neural circuits, and they are interpreted from an entirely neuronal perspective, i.e., they neglect the glial half of the brain altogether.

If we begin from the postulate that neurons produce thought, how might this work? There are two hypotheses; the first is quantitative and proposes that an increase in the number and complexity of neurons somehow gives rise to a new form of mental functioning called thought. The expansion of the neuronal network, with its increasingly intricate neuronal mesh, produces thought once its level of complexity surpasses a certain threshold. This is analogous to a locomotive steam engine boiler building up pressure. Once a certain pressure is achieved, the pistons are activated and the train is propelled forward. The second hypothesis is qualitative and relies on the observation that not all neurons are created equal. Some are smarter than others, like the multifaceted interneurons that are more involved with message integration than message transmission (Appendix IV).

In fact, there is no basis for choosing one neuronal hypothesis over another. The communication function of neurons does not exclude them from having additional as-yet-unidentified functions. One could even imagine that mental abilities do not arise from the rudimentary, yet well-coordinated, neural circuits; but rather from a diffuse and complex interplay involving many interconnected circuits, like the billions of molecules that collectively produce waves in the ocean.

What Role Do Blood Vessels Play in the "*Ménage à Trois*"?

Not to be overlooked are blood vessels, which, in the form of capillaries, not only nourish nerve cells and eliminate their waste, but also enable the brain's communication with the rest of the body.

The brain receives information on a continuous basis, not just from afferent nerves relaying stimuli from the skin and sensory organs, but also from chemical compounds, primarily hormones secreted by endocrine glands such as the thyroid and adrenal glands, that are transported via the circulatory system. Other blood-borne stimuli include chemical molecules found in the environment (oxygen, and a variety of nutrients) and various by-products produced by the body's organs, the intestines for example. Hence, capillaries are obligatory intermediaries between the body and the brain. In every instance, glial cells, and especially astrocytes, are the critical link between capillary blood flow and neuronal activity.

Blood vessels do not produce thought—at least not in ways we are currently aware of—but they do participate in brain functioning as part of the blood–brain barrier (see Figure 13). This barrier protects the "noble" brain, insulating it from the "plebeian" body. Here too, astrocytes get involved via their end feet, which ensure the transfer of nutrients from the blood to the neurons.

To recap, whereas blood vessels are essential to the functioning of the neuron–glial partnership, there is currently no evidence that they contribute directly to our mental functions. The brain and the body communicate using, among other channels, blood vessels as intermediaries. Although the body can modify thoughts via the blood—and also via afferent signals that emanate from a type of "interoceptive sensory system" that monitors and conveys information

on the state of our organ systems, notably the viscera—the body on
its own cannot create thoughts.

The Glial Cell: Another Candidate for the Title Role of Thought-Maker

Suppose scientists had never recorded neurons in the early 1900s?
Imagine instead that glial cell functions had become the focus of research, while neurons remained a mere sideshow. Today we would
most likely be discussing a model of brain function based on glial cell
physiology!

Glial Cells Are More Refined Interlocutors Than They Are Given Credit For!

We must distinguish between two types of glial cells, the oligodendrocytes that form the myelin sheath enveloping neuronal
axons (Appendix II) and the astrocytes that, through their tight
embrace of every synapse, essentially permeate the rest of the
nerve tissue space. Astrocytes may lack some attributes (the absence of electric currents running along their membranes, for example) but they possess other major assets, such as their ability
to carry on a continuous dialogue with neighboring neurons (see
Figures 17 and 18) and their capacity to propagate calcium waves,
by which they relay information across a vast swath of nervous
tissue (Figure 15).

From a biological perspective, astrocytes are not inert brain
building blocks. They manifest a biochemical potential that rivals or
even surpasses that of neurons. For example:

- *They are more sophisticated than neurons.* They make use of the
  same biochemical tools as neurons (neuro-/glio-transmitters,
  receptors, etc.) but also produce substances that neurons do not,

such as prostaglandins, certain peptides and amino acids, D-serine, ATP, lactate, etc.

- *They maintain an intimate relationship with neurons.* They control neurons and, in turn, are controlled by neurons. Thus, if neurons are involved in behavior, their astrocyte partners most certainly are too.
- *They have the extraordinary ability to communicate,* not only with neurons, but also among themselves via intercellular bridges called gap junctions.
- *By generating calcium waves* that are relatively long-lasting—an ability exclusive to them—they communicate with other astrocytes over long distances.
- *They act over a long period of time,* comparable to the timeframe over which behavioral changes such as learning and adaptation manifest.
- *They have the ability to maintain an environment* conducive to the proper functioning and well-being of neurons, as well as that of other nervous system cells.

Astrocytes are also formidable from the perspective of spatial organization:

- *They are abundant in brain regions where behaviors develop,* like the hippocampus—although, in itself, this does not prove their involvement in generating behaviors.
- *They present in an orderly distribution,* in the form of regular polyhedra. Their star-shape enables them to maintain contact with several hundred thousand synapses, as well as process selective information from the plethora of messages received from neurons.
- *They are assembled in a regular syncytium stack* and communicate with one another. Interestingly, the cells that facilitate communication throughout this syncytium are not neurons, but rather the astrocytes themselves.

Do Astrocytes Have More Reason to Think Than Do Neurons?

For the sake of simplicity, we mention only three of the many remarkable physiological features of astrocytes:

1. *They have the capacity to transmit messages*, albeit slowly, over long distances (several millimeters)—quite different from the very rapid transmission of electrical messages over long distances that is an exclusive feature of neurons.
2. *They have the capacity to integrate* electrical messages received from a great many surrounding nerve endings (see Figure 17).
3. *They possess the special feature of being able to synchronize neuronal* activity from varying sources, which makes it possible to shape a meaningful message from seemingly scattered and disparate information.

Given all this, the question remains, do astrocytes think? Imagine that were the case. Perhaps they form rudimentary thoughts that give rise to simple behaviors (for example, walking—a simple automatic behavior, or choosing to turn right rather than left), or even more complex thoughts like the abstract contemplation of the nature of consciousness. Do the known properties of astrocytes support the concept of "thinking" astrocytes? A review of the scientific literature does not allow us to go that far. Their capacity for transmission, integration, and synchronization is impressive, but does not countenance that sort of dialectical leap. So, at this point in time, a priori, we cannot proclaim that astrocytes think. In that case, what entity in the brain is engaged in thinking?

I am......
.....but do I think?

When we say "astrocyte," we are actually referring to an ensemble of astrocytes. Astrocytes do not function as isolated units; instead, their physiological properties reflect a collective effort. The same applies to neurons; we have no reason to believe that individual neurons are any more capable of thought than are individual astrocytes. What use would there be for intelligent yet isolated neurons that merely route physicochemical messages around in the absence of their intimate ties with the astrocytes that ensure their survival? What use would there be for intelligent yet isolated astrocytes that have no access to the external information provided by afferent neurons and no means of transmitting their mental musings in the absence of efferent neurons? Isn't it more likely that the neuron–astrocyte couple is the true unit needed to process sensory perceptions, elicit personal thoughts, and formulate plans that can adapt to changing circumstances?

Is the (Neuron + Astrocyte) Pair Superior to the Neuron + the Astrocyte?

Are neuron–astrocyte pairs more robust than individual neurons and astrocytes? In other words, are these cellular entities more effective when coupled? Just as with humans, the answer is complicated. Is a loving couple stronger than two single beings? In principle, yes— although Georges Feydau once quipped, "marriage is the art of two people living together as happy as they would have lived on their own." Analogous to human marriages, with which we are familiar, we speculate that neuron–astrocyte couples function more effectively than either cell type would alone. However, being more functional is one thing, creating a new substance is another. A human couple may have that ability, but does the same hold true for nerve cells?

Neurons and astrocytes form, in effect, couples that, along with blood vessels, constitute the emblematic *ménage à trois* of the brain's machinery (see Figure 17). Just as with some couples, opposites may attract. We might view neurons as being faster and more impulsive; whereas astrocytes, more cautious and deliberate, take on a nurturing role. They have different but compatible skills. They both engage in dialogue and they complement one another. Logically, the neuron–astrocyte union should be stronger than either cell type functioning in isolation. By joining forces do they create a new form of energy called "thought," which neither is capable of producing on its own? This remains to be proven.

We are, therefore,……
…….we think.

Can the "Non-thinking" Properties of the Neuron–Astrocyte Pair, Through a Sort of Miraculous Synergy, Produce a Transcendent Substance With Thinking Properties?

Do nerve cells create thought and, as a consequence, behaviors adapted to our circumstances? Why not? In the way that the chemical reaction between an acid and a base produces a new substance, i.e., a salt, can non-thinking matter interact to give rise to thinking non-matter? It could be; but in saying this we don't really answer the question at hand. In fact, we find ourselves back at the starting gate. This is because we have no clear idea of how to define the mental process we designate as thought in either mathematical or physical terms, and even less so in biological terms. Nonetheless, at the very least, it's clear that at this point in time we could benefit from some serious debunking of long-held concepts about mental functioning.

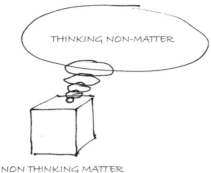

NON-THINKING MATTER

Conclusion

Addressing the Only Question Really Worth Asking: How Does the Brain Produce Thought?

If you are a believer in dualism, you see the mind and body as separate entities. You do not accept that the brain, a physical entity, is the source of thought, but rather that thought arises out of a nonphysical entity, the mind. Otherwise, you are likely a non-dualist (a monist) and you accept that mental processes are the product of the brain's physicochemical mechanisms, which produce thoughts either spontaneously or in response to the environment. Going forward, we have the option of adopting one of two perspectives. The first is that the mystery of thought creation has no solution. Although this fatalistic viewpoint has its devotees; it is really just a fallback to dualism and often manifests as an attitude that greets scientific advances with apathy or even sarcasm. The second perspective is that this enigma indeed has a solution—an intellectual stance that encourages one to think in terms of probabilities. This latter attitude helps to foster rapid-paced scientific progress. It could be that our ever-deepening understanding of the nervous system will inspire a solution sooner than expected, this despite the gap between the current state of neuroscientific knowledge and a more profound understanding of higher mental functions. Will some unexpected discovery shed light on the heretofore unidentified physicochemical properties of thought—who can say?

A paradox is that many researchers fear the unknown, i.e., they prefer to stay within their comfort zone. Many have made a name for themselves in their chosen field; so why not just rest on one's laurels? Funding will be easier and articles more readily accepted for publication. More to the point, we scientists fear losing what we have and there is no guarantee we will discover something new and exciting. How do we go about being creative and making discoveries without taking on too much risk? To be original, one must tread a new path. The first step is to review what we already know and be wary of what we believe in. It's easy to tell people to do the opposite of what they usually do. After all, we know whence we came, but not to where we are headed. We are generally wedded to our scientific routines. The option of just continuing down the well-trodden path is very appealing, especially when it is still productive. But we should also contemplate taking a detour. This sort of reflective, introspective process will prove essential in our quest to reveal the mechanisms underlying normal behaviors, as well as behaviors that deviate from normal. To forge a new outlook one must be audacious, particularly at a moment in time when the scientific literature still views mental functions and the behavior of living beings, including humans, from a largely neuronal perspective.

To challenge the conventional paradigm, we propose the involvement of glial cells in behavior; albeit not with a direct role—something for which we have no hard evidence. Rather, the work of glial cells is intertwined with that of neurons—the cells that relay information from the environment and transmit the impulses that trigger motor actions.

The enigma resides within the brain, the place where thought develops. "New results may suggest both new ideas and also alert us to errors in old conceptions," remarked Francis Crick. With this in mind, we note that research into neurons and behavior has been going on for nearly one hundred years, whereas research into astrocytes and behavior got underway a mere twenty years ago. With some trepidation, for fear of rocking the boat, we pose the obvious question: isn't it about time?

From this point forward, one of two possible scenarios could unfold. Either we humans are sufficiently adept, despite our finite intellectual capacities, and we will more or less solve this enigma down the road; or, lacking that ability, the solution will be left to more sophisticated beings—although, it must be said, we know of no benchmarks in this universe that let us assess our own mental capabilities. So, although it is a virtue to be humble and somewhat skeptical; nonetheless, let us forge ahead with ambition and *élan*!

Microglia

Microglia, another branch of the glial family, play a role in the central nervous system akin to that of certain white blood cells (macrophages) in other body tissues. When activated by an inflammatory process, they respond to immunological attacks and destroy microbes; however, they are a double-edged sword. Activated microglia cells not only destroy dangerous intruders but may also damage and kill neurons.

ASTROCYTE

MICROGLIAL CELL

Recent work also points to a physiological role for microglia in synaptic plasticity. During development, and even in into adulthood, microglia eliminate certain non-functional synaptic connections. In doing so, they help sculpt neuronal circuits so they operate more efficiently—yet another example of the close functional relationship between neurons and glial cells (Hong and Stevens, 2016).

Myelin: The Brain's White Matter

If you could peer into a brain that had been cut open, you would see two colors. Around the periphery (the cortex) and in the center (the basal ganglia) the brain matter appears pinkish gray, but between these two regions you would notice that the brain is bright white. Whereas the gray matter hosts the neuronal cell bodies with their nerve endings, the white matter contains the myelin that wraps around neuronal axons.

Myelin is a protective and insulating membrane sheath that surrounds many nerve fibers, akin to the plastic insulation that coats electrical wires (Zalc and Rosier, 2018). It is made up of oligodendrocytes that wrap themselves around neuronal axons. One function of myelin is to accelerate the small electric currents (nerve impulses) that propagate along axons. In non-myelinated nerve fibers these nerve impulses travel at speeds ranging from 0.5 to 2 m/s, but in myelinated fibers they can zoom up to 80–120 m/s, i.e., 300–400 km/h!

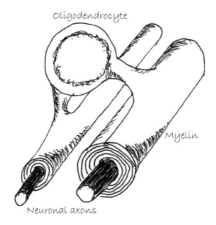

Let's look at an example that illustrates how important myelin is to the execution of behaviors. How long does it take a driver to brake a car to avoid an obstacle? The answer: about two one-hundredths of a second (0.02 s)! How does the nervous system so quickly cover the 2 m distance between the eye that sees the obstacle and the leg muscles that hit the brake? We can thank the myelin sheath wrapped around various nerve fibers that extend from the eye to the brain and then down to the legs. If these fibers were not myelinated, it would take up to 2 s to brake! You can imagine the implications for the car, the driver, and the obstacle!

Not all animals have myelin. Myelin came on to the scene more than 400 million years ago with our fish-like evolutionary ancestors. That is why nerve conduction in flies, an earlier-branching lineage with non-myelinated nerves, is comparatively slow. So why then is it so hard to catch a fly in your hands? Well, because the distance between the fly's eyes and its wing muscles is less than 2 cm; so even without myelin, the fly can take off in less than 0.02 s—fast enough to escape our grasp.

To sum up, myelin, a basic component of white matter, allows us to perform many actions with great speed. When myelin is damaged, for example in multiple sclerosis, nerve conduction no longer functions normally in those brain regions afflicted with white matter lesions or sclerotic plaques. Neurons may be partially denuded of their myelin sheath to the point where their survival is in question. An individual with multiple sclerosis may incur disabilities that progress in a chaotic succession of waves called flare-ups. Flare-ups can cause paralysis of a limb, loss of vision, an acute and sudden loss of skin sensation (paresthesia), urinary incontinence, and other symptoms.

Three Aspects of Brain Design

I: The human brain is the result of a multi-stage evolutionary process spanning several hundred million years. First a rudimentary nervous system gave rise to a spinal cord that transmitted information to nerves responsible for activating muscles (a). Then the basal ganglia developed (b), which enabled certain automatic behaviors such as walking. The most recent addition was the cerebral cortex (c), which controls non-automatic behaviors—a key feature in primates.

II: Information about sensory experiences (touch, vision, hearing, smell, taste) is dispatched to the brain's posterior cortex. Our motor responses (gestures, speech) are managed in the brain's frontal cortex.

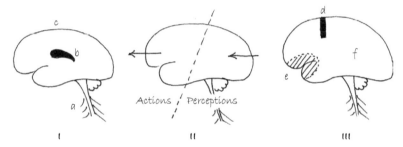

III: We can execute actions thanks to neurons that manage our motor functions (d), which send their commands to the spinal cord. Our emotions (joy, fear, disgust, etc., and their opposites) are processed and integrated in a region of the cerebral cortex called the limbic system (e). Finally, our intellect (discernment, insight, memory, language, etc.) is conceived and controlled in the other regions of the cerebral cortex called the association cortex (f).

Interneurons and Neurons that Modulate other Neurons

There is a type of neuron called an interneuron that does not convey information remotely. Instead it helps other neurons work together harmoniously. Interneurons act as synchronizers, i.e., they integrate the activity of several hundred neurons so that the whole group delivers a coherent message.

About 20% of the neurons in the cerebral cortex are interneurons and their axons project locally within the cortex. There are dozens of types of interneurons with evocative names like chandelier, double bouquet, bitufted, and bipolar, and they were all described by Ramón y Cajal through his morphological observations. "Like the entomologist hunting for brightly colored butterflies," he wrote in 1923, "my attention was drawn to the flower garden of the gray matter, which contained cells with delicate and elegant forms, the mysterious butterflies of the soul, the beating of whose wings may someday . . . clarify the secrets of mental life." (Ramón y Cajal, 1923).

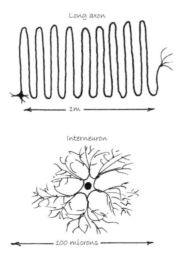

The interneurons utilize the inhibitory neurotransmitter gamma-aminobutyric acid to communicate with other neurons. They also employ a second neurotransmitter (usually a peptide such as vasoactive intestinal peptide) to speak to astrocytes and capillaries (Magistretti and Allaman, 2015), setting up, in effect, a *ménage à trois*.

Neurons that modulate other neurons are different from neurons that engage in point-to-point communication, in that they do not directly and selectively influence the activity of the neurons they project to. Rather they enable these neurons to remain in an active state, or not. This activating effect persists far longer than a few milliseconds, which means they can modulate the behavior of neuronal groups on a long-term basis. This prolonged timeframe is consistent with certain aspects of human behavior, such as learning, where the process unfolds over an extended period of time. So, it seems that by promoting more efficient integration of neuronal activity, neuromodulation lays the groundwork for the successful execution of complex behaviors.

One example of a modulatory system is the neural circuit originating in the brainstem nucleus called the locus cœruleus. The neuronal cell bodies are located in the locus cœruleus but their axons project throughout the brain and release the neurotransmitter noradrenaline all along their length—like an irrigation hose irrigating a lawn. This noradrenergic (noradrenalin-utilizing) system plays an important role in attention-related phenomena (Aston-Jones and Waterhouse, 2016).

The Global Neuronal Workspace

The global neuronal workspace model of consciousness (Dehaene et al., 1998) asserts that our brain processes information by means of interconnected cerebral networks, in a manner analogous to a mini-Internet. The Internet resembles a Russian nesting doll, layer enclosed within layer, where individual home-based or workplace computers connect to a multitude of local networks that address specific issues. Then these local networks report up a hierarchy to secondary networks. In turn, the secondary networks integrate all the information from the local networks and report it across the World Wide Web. In the global neuronal workspace model, the brain's localized cerebral networks process information at an unconscious level. The subjective experience of consciousness arises only once this information is amplified and made available "globally" throughout the brain.

A word of caution here regarding *brain-as-computer* comparisons. Although the human brain programs the computer there are salient differences between the two. Computers can "remember" huge amounts of data, whereas human brains have a limited memory capacity. Our brains, however, have a distinct advantage over computers; their immense number of nerve connections and their ability to process information is seemingly infinite. What's more, our brains' response patterns are not fixed and our nerve connections undergo constant remodeling. One might say "we never use the same brain twice" (Ansermet and Magistretti, 2004).

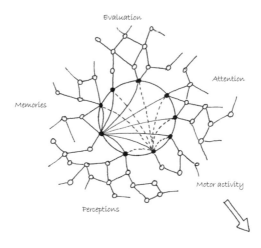

Nerve Cells Born from Stem Cells

As our nervous system develops, our stem cells differentiate into two cell groups: neurons and glial cells. The latter subdivide into astrocytes and oligodendrocytes.

A stem cell has two basic features. First, unlike neurons, stem cells renew themselves through cellular division. Theoretically they can divide indefinitely—a sort of self-sustaining renewal process. They are also "multipotent," in that they give birth to multiple types of nerve cells.

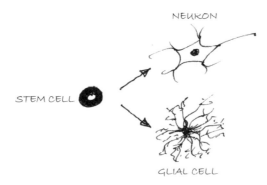

These dividing multipotent stem cells renew themselves in two different ways. Either they produce two identical daughter cells that follow in the footsteps of the mother cell (symmetric division) or they produce one daughter cell identical to the mother cell and one mature daughter cell that gives rise to other cell types (asymmetric division).

List of Figures

References

Agid Y., *L'Homme subconscient. Le cerveau et ses erreurs*, Paris, Robert Laffont, 2013.

Aimone J. B., Li Y., Lee S. W., Clemenson G. D., Deng W., Gage F. H., "Regulation and function of adult neurogenesis: From genes to cognition," *Physiol. Rev.*, 2014, 94 (4): 991–1026.

Allaman I., Belanger M., Magistretti P., "Astrocyte–neuron metabolic relationships: For better and for worse," *Trends Neurosci.*, 2011, 34 (2): 76–87.

Allaman I., Gavillet M., Belanger M., Laroche T., Viertl D., Lashuel H. A., Magistretti P., "Amyloid-beta aggregates cause alterations of astrocytic metabolic phenotype: Impact on neuronal viability," *J. Neurosci.*, 2010, 30 (9): 3326–3338.

Anderson M. A., O'Shea T. M., Burda J. E., Ao Y., Barlatey S. L., Bernstein A. M., Kim J. H., James N. D., Rogers A., Kato B., Wollenberg A. L., Kawaguchi R., Coppola G., Wang C., Deming T. J., He Z., Courtine G., Sofroniew M. V., "Required growth facilitators propel axon regeneration across complete spinal cord injury," *Nature*, 2018, 561 (7723): 396–400.

Andriezen W. L., "On a system of fibre-cells surrounding the blood-vessels of the brain of man and mammals, and its physiological significance," *Internationale Monatsschrift für Anatomie und Physiologie*, 1893, 19: 532–540.

Ansermet F., Magistretti P., *À chacun son cerveau. Plasticité neuronale et inconscient*, Paris, Odile Jacob, 2004.

Araque A., Carmignoto G., Haydon P. G., Oliet S. H., Robitaille R., Volterra A., "Gliotransmitters travel in time and space," *Neuron*, 2014, 81 (4): 728–739.

Aston-Jones G., Waterhouse B., "Locus coeruleus: From global projection system to adaptive regulation of behavior," *Brain Res.*, 2016, 1645: 75–78.

Barres B. A., "The mystery and magic of glia: A perspective on their roles in health and disease," *Neuron*, 2008, 60 (3): 430–440.

Barreto G., White R. E., Ouyang Y., Xu L., Giffard R. G., "Astrocytes: Targets for neuroprotection in stroke," *Cent. Nerv. Syst. Agents Med. Chem.*, 2011, 11 (2): 164–173.

Bellesi M., de Vivo L., Tononi G., Cirelli C., "Effects of sleep and wake on astrocytes: Clues from molecular and ultrastructural studies," *BMC Biol.*, 2015, 13: 66.

Bentivoglio M., Mazzarello P., "The history of radial glia," *Brain Res. Bull.*, 1999, 49 (5): 305–315.

Bernardinelli Y., Magistretti P., Chatton J.-Y., "Astrocytes generate Na+-mediated metabolic waves," *Proc. Natl Acad. Sci. USA*, 2004, 101 (41): 14937–14942.

Bezzi P., Carmignoto G., Pasti L., Vesce S., Rossi D., Rizzin B. L., Pozzan T., Volterra A., "Prostaglandins stimulate calcium-dependent glutamate release in astrocytes," *Nature*, 1998, 391 (6664): 281–285.

Black J. E., Isaacs K. R., Anderson B. J., Alcantara A. A., Greenough T., "Learning causes synaptogenesis, whereas motor-activity causes angiogenesis, in cerebellar cortex of adult-rats," *Proc. Natl Acad. Sci. USA*, 1990, 87 (14): 5568–5572.

Boury-Jamot B., Carrard A., Martin J. L., Halfon O., Magistretti P., Boutrel B., "Disrupting astrocyte–neuron lactate transfer persistently reduces conditioned responses to cocaine," *Mol. Psychiatry*, 2016, 21 (8): 1070–1076.

Bushong E. A., Martone M. E., Jones Y. Z., Ellisman M. H., "Protoplasmic astrocytes in CA1 stratum radiatum occupy separate anatomical domains," *J. Neurosci.*, 2002, 22 (1): 183–192.

Cali C., Baghabra J., Boges D. J., Holst G. R., Kreshuk A., Hamprecht F. A., Srinivasan M., Lehvaslaiho H., Magistretti P., "Three-dimensional immersive virtual reality for studying cellular compartments in 3D models from EM preparations of neural tissues," *J. Comp. Neurol.*, 2016, 524 (1): 23–38.

Cao X., Li L. P., Wang Q., Wu Q., Hu H. H., Zhang M., Fang Y. Y., Zhang J., Li S. J., Xiong W. C., Yan H. C., Gao Y. B., Liu J. H., Li W., Sun L. R., Zeng Y. N., Zhu X. H., Gao T. M., "Astrocyte-derived ATP modulates depressive-like behaviors," *Nat. Med.*, 2013, 19 (6): 773–777.

Carrard A., Elsayed M., Margineanu M., Boury-Jamot B., Fragnière L., Meylan E. M., Petit J. M., Fiumelli H., Magistretti P. J., Martin J. L., "Peripheral administration of lactate produces antidepressant-like effects," *Mol. Psychiatry*, 2018, (2): 392–399.

Changeux J.-P., *L'Homme neuronal*, Paris, Fayard, 1983.

Charles A., "Glia–neuron intercellular calcium signaling," *Dev. Neurosci.*, 1994, 16 (3–4): 196–206.

Charles A., "Intercellular calcium waves in glia," *Glia*, 1998, 24 (1): 39–49.

Charles A., "Reaching out beyond the synapse: Glial intercellular waves coordinate metabolism," *Sci. STKE*, 2005, 270: pe6.

Chotard C., Salecker I., "Neurons and glia: Team players in axon guidance," *Trends Neurosci.*, 2004, 27 (11): 655–661.

Churchill M. J., Wesselingh S. L., Cowley D., Pardo C. A., McArthur J. C., Brew B. J., Gorry P. R., "Extensive astrocyte infection is prominent in human immunodeficiency virus-associated dementia," *Ann. Neurol.*, 2009, 66 (2): 253–258.

Clapham D. E., "Calcium signaling," *Cell*, 1995, 80 (2): 259–268.

Clarke L. E., Barres B. A., "Emerging roles of astrocytes in neural circuit development," *Nat. Rev. Neurosci.*, 2013, 14 (5): 311–321.

Cleveland D. W., Rothstein J. D., "From Charcot to Lou Gehrig: Deciphering selective motor neuron death in ALS," *Nat. Rev. Neurosci.*, 2001, 2 (11): 806–819.

Cornell-Bell A. H., Finkbeiner S. M., Cooper M. S., Smith S. J., "Glutamate induces calcium waves in cultured astrocytes: long-range glial signaling," *Science*, 1990, 247 (4941): 470–473.

Damier P., Hirsch E. C., Zhang P., Agid Y., Javoy-Agid F., "Glutathione peroxidase, glial cells and Parkinson's disease," *Neuroscience*, 1993, 52 (1): 1–6.

Danbolt N. C., "Glutamate uptake," *Prog. Neurobiol.*, 2001, 65 (1): 1–105.

DeFelipe J., *Cajal's Butterflies of the Soul*, Oxford, Oxford University Press, "Science and Art," 2010.

Dehaene S., *Le Code de la conscience*, Paris, Odile Jacob, 2014.

Dehaene S., Kerszberg M., Changeux J.-P., "A neuronal model of a global workspace in effortful cognitive tasks," *Proc. Natl Acad. Sci. USA*, 1998, 95 (24): 14529–14534.

De Pitta M., Brunel N., Volterra A., "Astrocytes: Orchestrating synaptic plasticity?," *Neuroscience*, 2016, 323: 43–61.

Diamond M. C., Scheibel A. B., Jr Murphy G. M., Harvey T., "On the brain of a scientist: Albert Einstein," *Exp. Neurol.*, 1985, 88 (1): 198–204.

Dimou L., Gotz M., "Glial cells as progenitors and stem cells: New roles in the healthy and diseased brain," *Physiol. Rev.*, 2014, 94 (3): 709–737.

Dringen R., Gutterer J. M., Hirrlinger J., "Glutathione metabolism in brain metabolic interaction between astrocytes and neurons in the defense against reactive oxygen species," *Eur. J. Biochem.*, 2000, 267 (16): 4912–4916.

Duncan J. S., Sander J. W., Sisodiya S. M., Walker M. C., "Adult epilepsy," *Lancet*, 2006, 367 (9516): 1087–1100.

Edelman G. M., *Biologie de la conscience*, Paris, Odile Jacob, 1992.

Egeland M., Zunszain P. A., Pariante C. M., "Molecular mechanisms in the regulation of adult neurogenesis during stress," *Nat. Rev. Neurosci.*, 2015, 16 (4): 189–200.

Elsayed M., Magistretti P., "A new outlook on mental illnesses: Glial involvement beyond the glue," *Front. Cell. Neurosci.*, 2015, 9: 468.

Emery P., Freeman M. R., "Glia got rhythm," *Neuron*, 2007, 55 (3): 337–339.

Eriksson P. S., Perfilieva E., Björk-Eriksson T., Alborn A. M., Nordborg C., Peterson D. A., Gage F. H., "Neurogenesis in the adult human hippocampus," *Nat. Med.*, 1998, 4 (11): 1313–1317.

Eroglu C., Barres B. A., "Regulation of synaptic connectivity by glia," *Nature*, 2010, 468 (7321): 223–231.

Fellin T., Ellenbogen J. M., De Pitta M., Ben-Jacob E., Halassa M. M., "Astrocyte regulation of sleep circuits: Experimental and modeling perspectives," *Front. Comput. Neurosci.*, 2012, 6: 65.

Ferraiuolo L., Higginbottom A., Heath P. R., Barber S., Greenald D., Kirby J., Shaw P. J., "Dysregulation of astrocyte–motoneuron cross-talk in mutant superoxide dismutase 1-related amyotrophic lateral sclerosis," *Brain*, 2011, 134 (9): 2627–2641.

Finsterwald C., Magistretti P., Lengacher S., "Astrocytes: New targets for the treatment of neurodegenerative diseases," *Curr. Pharm. Des.*, 2015, 21 (25): 3570–3581.

Forman M. S., Lal D., Zhang B., Dabir D. V., Swanson E., Lee V. M., Trojanowski Q., "Transgenic mouse model of tau pathology in astrocytes leading to nervous system degeneration," *J. Neurosci.*, 2005, 25 (14): 3539–3550.

Fu W., Jhamandas J. H., "Role of astrocytic glycolytic metabolism in Alzheimer's disease pathogenesis," *Biogerontology*, 2014, 15 (6): 579–586.

Golgi C., "Sulla struttura della sostanza grigia del cervello," *Gazzetta medica italiana Lombardia*, 1873, 33: 244–246.

Gould E., Tanapat P., "Stress and hippocampal neurogenesis," *Biol. Psychiatry*, 1999, 46 (11): 1472–1479.

Grosjean Y., Grillet M., Augustin H., Ferveur J. F., Featherstone D. E., "A glial amino-acid transporter controls synapse strength and courtship in *Drosophila*," *Nat. Neurosci.*, 2008, 11 (1): 54–61.

Gundersen V., Storm-Mathisen J., Bergersen L. H., "Neuroglial transmission," *Physiol. Rev.*, 2015, 95 (3): 695–726.

Gupta A., Tsai L. H., Wynshaw-Boris A., "Life is a journey: A genetic look at neocortical development," *Nature Rev. Genet.*, 2002, 3 (5): 342–355.

Halassa M. M., Fellin T., Haydon P. G., "The tripartite synapse: Roles for gliotransmission in health and disease," *Trends Mol. Med.*, 2007, 13 (2): 54–63.

Halassa M. M., Fellin T., Takano H., Dong J. H., Haydon P. G., "Synaptic islands defined by the territory of a single astrocyte," *J. Neurosci.*, 2007, 27 (24): 6473–6477.

Han X., Chen M., Wang F., Windrem M., Wang S., Shanz S., et al., "Forebrain engraftment by human glial progenitor cells enhances synaptic plasticity and learning in adult mice," *Cell Stem Cell*, 2013, 12 (3): 342–353.

Haydon P. G., Nedergaard M., "How do astrocytes participate in neural plasticity?", *Cold Spring Harbor Perspect. Biol.*, 2015, 7 (3): a020438.

Henneberger C., Papouin T., Oliet S. H., Rusakov D. A., "Long-term potentiation depends on release of D-serine from astrocytes," *Nature*, 2010, 463 (7278): 232–236.

Herculano-Houzel S., "The human brain in numbers: A linearly scaled-up primate brain," *Front. Hum. Neurosci.*, 2009, 3: 31.

Herculano-Houzel S., "The glia/neuron ratio: How it varies uniformly across brain structures and species and what that means for brain physiology and evolution," *Glia*, 2014, 62 (9): 1377–1391.

Hong S., Stevens B., "Microglia: Phagocytosing to clear, sculpt, and eliminate," *Dev. Cell.*, 2016, 38 (2): 126–128.

Howarth C., "The contribution of astrocytes to the regulation of cerebral blood flow," *Front. Neurosci.*, 2014, 8: 103.

Jourdain P., Bergersen L. H., Bhaukaurally K., Bezzi P., Santello M., Domercq M., Matute C., Tonello F., Gundersen V., Volterra A., "Glutamate exocytosis from astrocytes controls synaptic strength," *Nat. Neurosci.*, 2007, 10 (3): 331–339.

Jourdain P., Allaman I., Rothenfusser K., Fiumelli H., Marquet P., Magistretti P., "L-Lactate protects neurons against excitotoxicity: Implication of an ATP-mediated signaling cascade," *Sci. Rep.*, 2016, 6: 21250.

Jouvet M., "Neurophysiology of the states of sleep," *Physiol. Rev.*, 1967, 47 (2): 117–177.

Jouvet M., Le Château des songes, Paris, Odile Jacob, 1992.

Kato T. A., Watabe M., Kanba S., "Neuron–glia interaction as a possible glue to translate the mind–brain gap: A novel multi-dimensional approach toward psychology and psychiatry," *Front. Psychiatry*, 2013, 4: 139.

Kempermann G., Song H., Gage F. H., "Neurogenesis in the adult hippocampus," *Cold Spring Harbor Perspect. Biol.*, 2015, 7 (9): a018812.

Kettenmann H., Verkhratsky A., "Neuroglia: The 150 years after," *Trends Neurosci.*, 2008, 31 (12): 653–659.

Kimelberg H. K., "Astrocytic swelling in cerebral ischemia as a possible cause of injury and target for therapy," *Glia*, 2005, 50 (4): 389–397.

Kleim J. A., Markham J. A., Vij K., Freese J. L., Ballard D. H., Greenough W. T., "Motor learning induces astrocytic hypertrophy in the cerebellar cortex," *Behav. Brain Res.*, 2007, 178 (2): 244–249.

Lee Y., Morrison B. M., Li Y., Lengacher S., Farah M. H., Hoffman P. N., Liu Y., Tsingalia A., Jin L., Zhang P. W., Pellerin L., Magistretti P., Rothstein

J. D., "Oligodendroglia metabolically support axons and contribute to neurodegeneration," *Nature*, 2012, 487 (7408): 443–448.

Le Meur K., Mendizabal-Zubiaga J., Grandes P., Audinat E., "GABA release by hippocampal astrocytes," *Front. Comput. Neurosci.*, 2012, 6: 59.

Liu C. C., Takahisa K., Huaxi X., Guojun Bu B., "Apolipoprotein E and Alzheimer disease: Risk, mechanisms, and therapy," *Nat. Rev. Neurol.*, 2013, 9 (2): 106–118.

Liu X., Gangoso E., Yi C., Jeanson T., Kandelman S., Mantz J., Giaume C., "General anesthetics have differential inhibitory effects on gap junction channels and hemichannels in astrocytes and neurons," *Glia*, 2016, 64 (4): 524–536.

Lizardi-Cervera J., Almeda P., Guevara L., Uribe M., "Hepatic encephalopathy: A review," *Ann. Hepatol.*, 2003, 2 (3): 122–130.

Llinas R. R., Pare D., "Of dreaming and wakefulness," *Neuroscience*, 1991, 44 (3): 521–535.

Lobsiger C. S., Cleveland D. W., "Glial cells as intrinsic components of non-cell-autonomous neurodegenerative disease," *Nat. Neurosci.*, 2007, 10 (11): 1355–1360.

Lugaro E., "Sulk funzioni della nevroglia," *Riv. Pat. Nerv. Ment.*, 1907, 12: 225–233.

Machler P., Wyss M. T., Elsayed M., Stobart J., Gutierrez R., von Faber-Castell A., Kaelin V., Zuend M., San Martin A., Romero-Gomez I., Baeza-Lehnert F., Lengacher S., Schneider B. L., Aebischer P., Magistretti P., Barros L. F., Weber B., "In vivo evidence for a lactate gradient from astrocytes to neurons," *Cell. Metab.*, 2016, 23 (1): 94–102.

Magistretti P., Allaman I., "A cellular perspective on brain energy metabolism and functional imaging," *Neuron*, 2015, 86 (4): 883–901.

Magistretti P., Allaman I., "Lactate in the brain: from metabolic end-product to signaling molecule," *Nat. Rev. Neurosci.*, 2018, 19: 235–249.

Magistretti P., Chatton J.-Y., "Relationship between L-glutamate-regulated intracellular (Na⁺) dynamics and ATP hydrolysis in astrocytes," *J. Neural. Transm.*, 2005, 112 (1): 77–85.

Magistretti P., Manthorpe M., Bloom F. E., Varon S., "Functional receptors for vasoactive intestinal polypeptide in cultured astroglia from neonatal rat brain," *Regul. Pept.*, 1983, 6 (1): 71–80.

Mantz J., Cordier J., Giaume C., "Effects of general anesthetics on intercellular communications mediated by gap junctions between astrocytes in primary culture," *Anesthesiology*, 1993, 78 (5): 892–901.

Marinesco M. G., "Lesions des centres nerveux produites par la toxine du *Bacillus Botulinus*," *C. R. Soc. Biol. (Paris)*, 1896, 48: 989–991.

Markham J. A., Greenough W. T., "Experience-driven plasticity: beyond the synapse," *Neuron Glia Biol.*, 2004, 1 (4): 351–363.

Merkle F. T., Tramontin A. D., Garcia-Verdugo J. M., Alvarez-Buylla A., "Radial glia give rise to adult neural stem cells in the subventricular zone," *Proc. Natl Acad. Sci. USA*, 2004, 101 (50): 17528–17532.

Messing A., Brenner M., Feany M. B., Nedergaard M., Goldman J. E., "Alexander disease," *J. Neurosci.*, 2012, 32 (15): 5017–5023.

Miller G., "Neuroscience. The dark side of glia," *Science*, 2005, 308 (5723): 778–781.

Mirescu C., Gould E., "Stress and adult neurogenesis," *Hippocampus*, 2006, 16 (3): 233–238.

Mori T., Buffo A., Götz M., "The novel roles of glial cells revisited: The contribution of radial glia and astrocytes to neurogenesis," *Curr. Top. Dev. Biol.*, 2005, 69: 67–99.

Morris R. G. M., Oertel W., Gaebel W., Goodwin G. M., Little A., Montellano P., Westphal M., Nutt D. J., Di Luca M., "Consensus statement on European brain research: The need to expand brain research in Europe – 2015," *Eur. J. Neurosci.*, 2016, 44 (3): 1919–1926.

Nedergaard M., Ransom B., Goldman S. A., "New roles for astrocytes: Redefining the functional architecture of the brain," *Trends Neurosci.*, 2003, 26 (10): 523–530.

Newman E. A., "New roles for astrocytes: Regulation of synaptic transmission," *Trends Neurosci.*, 2003, 26 (10): 536–542.

Oberheim N. A., Wang X., Goldman S., Nedergaard M., "Astrocytic complexity distinguishes the human brain," *Trends Neurosci.*, 2006, 29 (10): 547–553.

Oberheim N. A., Takano T., Han X., He W., Lin J. H., Wang F., Xu, Wyatt J. D., Pilcher W., Ojemann J. G., Ransom B. R., Goldman S. A., Nedergaard M., "Uniquely hominid features of adult human astrocytes," *J. Neurosci.*, 2009, 29 (10): 3276–3287.

Oliet S. H., Mothet J. P., "Molecular determinants of D-serine-mediated gliotransmission: From release to function," *Glia*, 2006, 54 (7): 726–737.

Olney J. W., "Brain lesions, obesity, and other disturbances in mice treated with monosodium glutamate," *Science*, 1969, 164 (3880): 719–721.

Pannasch U., Rouach N., "Emerging role for astroglial networks in information processing: From synapse to behavior," *Trends Neurosci.*, 2013, 36 (7): 405–417.

Pannasch U., Vargová L., Reingruber J., Ezan P., Holcman D., Giaume C., Syková E., Rouach N., "Astroglial networks scale synaptic activity and plasticity," *Proc. Natl Acad. Sci. USA*, 2011, 108 (20): 8467–8472.

Papouin T., Ladepeche L., Ruel J., Sacchi S., Labasque M., Hanini M., Groc L., Pollegioni L., Mothet J. P., Oliet S. H., "Synaptic and extrasynaptic NMDA receptors are gated by different endogenous coagonists," *Cell*, 2012, 150 (3): 633–646.

Parpura V., Basarsky T. A., Liu F., Jeftinija K., Jeftinija S., Haydon P. G., "Glutamate-mediated astrocyte–neuron signalling," *Nature*, 1994, 369 (6483): 744–747.

Pellerin L., Magistretti P., "Glutamate uptake into astrocytes stimulates aerobic glycolysis: A mechanism coupling neuronal activity to glucose utilization," *Proc. Natl Acad. Sci. USA*, 1994, 91 (22): 10625–10629.

Pellerin L., Magistretti P., "Glutamate uptake stimulates Na+/K+-ATPase activity in astrocytes via activation of a distinct subunit highly sensitive to ouabain," *J. Neurochem.*, 1997, 69 (5): 2132–2137.

Perea G., Navarrete M., Araque A., "Tripartite synapses: Astrocytes process and control synaptic information," *Trends Neurosci.*, 2009, 32 (8): 421–431.

Petit J.-M., Gyger J., Burlet-Godinot S., Fiumelli H., Martin J. L., Magistretti P., "Genes involved in the astrocyte–neuron lactate shuttle (ANLS) are specifically regulated in cortical astrocytes following sleep deprivation in mice," *Sleep*, 2013, 36 (10): 1445–1458.

Pham K., Nacher J., Hof P. R., McEwen B. S., "Repeated restraint stress suppresses neurogenesis and induces biphasic PSA-NCAM expression in the adult rat dentate gyrus," *Eur. J. Neurosci.*, 2003, 17 (4): 879–886.

Philips T., Rothstein J. D., "Glial cells in amyotrophic lateral sclerosis," *Exp. Neurol.*, 2014, 262 (Pt B): 111–120.

Radford R. A., Morsch M., Rayner S. L., Cole N. J., Pountney D. L., Chung R. S., "The established and emerging roles of astrocytes and microglia in amyotrophic lateral sclerosis and frontotemporal dementia," *Front. Cell. Neurosci.*, 2015, 9: 414.

Rajkowska G., Stockmeier C. A., "Astrocyte pathology in major depressive disorder: Insights from human postmortem brain tissue," *Curr. Drug Targets*, 2013, 14 (11): 1225–1236.

Rakic P., "Elusive radial glial cells: historical and evolutionary perspective," *Glia*, 2003, 43 (1): 19–32.

Ramón y Cajal S., "Contribución al conocimiento de la neuroglia del cerebro humano [Contribution to the knowledge of neuroglia in the human brain] ," *Trabajos del Laboratorio de Investigaciones biológicas*, 1913, 11: 255–315.

Ramón y Cajal S., *Recuerdos de mi vida*, Madrid, Imprenta de Juan Pueyo, 1923. English translation: *Recollections of my Life*, Cambridge (MA), MIT Press, 1966.

Ramón y Cajal S. *Neuronismo o reticularismo? Las pruebas objeticas de la unitad anatomica de las celulas nerviosas*, Madrid, Consejo Superior de Investigaciones Scientificas, Instituto Ramón y Cajal, 1933. English translation: *Neuron Theory or Reticular Theory? Objective Evidence of the Anatomical Unity of Nerve Cells*, Madrid, Consejo Superior de Investigaciones Científicas, Instituto Ramón y Cajal, 1954.

Río Hortega P. del, "La microglia y su transformación en celulas en basoncito y cuerpos granulo-adiposos," *Trab. Lab. Invest. Biol., Madrid*, 1920, 18: 37–82.

Robertson J. M., "The astrocentric hypothesis: Proposed role of astrocytes in consciousness and memory formation," *J. Physiol. Paris*, 2002, 96 (3–4): 251–255.

Sada N., Lee S., Katsu T., Otsuki T., Inoue T., "Epilepsy treatment. Targeting LDH enzymes with a stiripentol analog to treat epilepsy," *Science*, 2015, 347 (6228): 1362–1367.

Sanacora G., Banasr M., "From pathophysiology to novel antidepressant drugs: Glial contributions to the pathology and treatment of mood disorders," *Biol. Psychiatry*, 2013, 73 (12): 1172–1179.

Santarelli L., Saxe M., Gross C., Surget A., Battaglia F., Dulawa S., Weisstaub N., Lee J., Duman R., Arancio O., Belzung C., Hen R., "Requirement of hippocampal neurogenesis for the behavioral effects of antidepressants," *Science*, 2003, 301 (5634): 805–809.

Sattler R., Tymianski M., "Molecular mechanisms of glutamate receptor-mediated excitotoxic neuronal cell death," *Mol. Neurobiol.*, 2001, 24 (1–3): 107–129.

Schummers J., Yu H., Sur M., "Tuned responses of astrocytes and their influence on hemodynamic signals in the visual cortex," *Science*, 2008, 320 (5883): 1638–1643.

Seifert G., Schilling K., Steinhauser C., "Astrocyte dysfunction in neurological disorders: A molecular perspective," *Nat. Rev. Neurosci.*, 2006, 7 (3): 194–206.

Sild M., Ruthazer E. S., "Radial glia: Progenitor, pathway, and partner," *Neuroscientist*, 2011, 17 (3): 288–302.

Silver J., Miller J. H., "Regeneration beyond the glial scar," *Nat. Rev. Neurosci.*, 2004, 5 (2): 146–156.

Singh A., Tetreault L., Kalsi-Ryan S., Nouri A., Fehlings M. G., "Global prevalence and incidence of traumatic spinal cord injury," *Clin. Epidemiol.*, 2014, 6: 309–331.

Sirevaag A. M., Greenough W. T., "Differential rearing effects on rat visual cortex synapses. III. Neuronal and glial nuclei, boutons, dendrites, and capillaries," *Brain Res.*, 1987, 424: 320–332.

Sirevaag A. M., Greenough W. T., "Plasticity of Gfap-immunoreactive astrocyte size and number in visual-cortex of rats reared in complex environments," *Brain Res.*, 1991, 540 (1–2): 273–278.

Slezak M., Pfrieger F. W., "New roles for astrocytes: Regulation of CNS synaptogenesis," *Trends Neurosci.*, 2003, 26 (10): 531–535.

Sofroniew M. V., Vinters H. V., "Astrocytes: Biology and pathology," *Acta Neuropathol.*, 2010, 119 (1): 7–35.

Somjen G. G., "Nervenkitt: Notes on the history of the concept of neuroglia," *Glia*, 1988, 1 (1): 2–9.

Sultan S., Li L., Moss J., Petrelli F., Cassé F., Gebara E., Lopatar J., Pfrieger F. W., Bezzi P., Bischofberger J., Toni N., "Synaptic integration of adult-born hippocampal neurons is locally controlled by astrocytes," *Neuron*, 2015, 88 (5): 957–972.

Suzuki A., Stern S. A., Bozdagi O., Huntley G. W., Walker R. H., Magistretti P., Alberini C. M., "Astrocyte–neuron lactate transport is required for long-term memory formation," *Cell*, 2011, 144 (5): 810–823.

Takano T., Tian G. F., Peng W., Lou N., Libionka W., Han X., Nedergaard M., "Astrocyte-mediated control of cerebral blood flow," *Nat. Neurosci.*, 2006, 9 (2): 260–267.

Theis M., Giaume C., "Connexin-based intercellular communication and astrocyte heterogeneity," *Brain Res.*, 2012, 1487: 88–98.

Theis M., Jauch R., Zhuo L., Speidel D., Wallraff A., Doring B., Frisch C., Sohl G., Teubner B., Euwens C., Huston J., Steinhauser C., Messing A., Heinemann U., Willecke K., "Accelerated hippocampal spreading depression and enhanced locomotory activity in mice with astrocyte-directed inactivation of connexin43," *J. Neurosci.*, 2003, 23 (3): 766–776.

Tian G. F., Azmi H., Takano T., Xu Q., Peng W., Lin J., Oberheim N., Lou N., Wang X., Zielke H. R., Kang J., Nedergaard M., "An astrocytic basis of epilepsy," *Nat. Med.*, 2005, 11 (9): 973–981.

Tononi G., Cirelli C., "Sleep and the price of plasticity: From synaptic and cellular homeostasis to memory consolidation and integration," *Neuron*, 2014, 81 (1): 12–34.

Tso M. C., Herzog E. D., "Was Cajal right about sleep?", *BMC Biology*, 2015, 13: 67.

Verkhratsky A., Butt A., *Glial Physioloy and Pathophysiology*, Hoboken (NJ), Wiley–Blackwell, 2013.

Volterra A., Meldolesi J., "Astrocytes, from brain glue to communication elements: The revolution continues," *Nat. Rev. Neurosci.*, 2005, 6 (8): 626–640.

Voutsinos-Porche B., Bonvento G., Tanaka K., Steiner P., Welker E., Chatton J.-Y., Magistretti P., Pellerin L., "Glial glutamate transporters mediate a functional metabolic crosstalk between neurons and astrocytes in the mouse developing cortex," *Neuron*, 2003, 37 (2): 275–286.

Warner-Schmidt J. L., Duman R. S., "Hippocampal neurogenesis: Opposing effects of stress and antidepressant treatment," *Hippocampus*, 2006, 16 (3): 239–249.

Weber B., Keller A. L., Reichold J., Logothetis N. K., "The microvascular system of the striate and extrastriate visual cortex of the macaque," *Cereb. Cortex*, 2008, 18 (10): 2318–2330.

Wentlandt K., Samoilova M., Carlen P. L., El Beheiry H., "General anesthetics inhibit gap junction communication in cultured organotypic hippocampal slices," *Anesth. Analg.*, 2006, 102 (6): 1692–1698.

World Health Organization, *Epilepsy: A Public Health Imperative: Summary*, 2019.

Yang J., Ruchti E., Petit J.-M., Jourdain P., Grenningloh G., Allaman I., Magistretti P., "Lactate promotes plasticity gene expression by potentiating NMDA signaling in neurons," *Proc. Natl Acad. Sci. USA*, 2014, 111 (33): 12228–12233.

Zalc B., Rosier F., *Myelin, The Brain's Supercharger*, New York, Oxford University Press, 2018.

Zhang Y., Barres B. A., "Astrocyte heterogeneity: An underappreciated topic in neurobiology," *Curr. Opin. Neurobiol.*, 2010, 20 (5): 588–594.

Zhao Y., Rempe D. A., "Targeting astrocytes for stroke therapy," *Neurotherapeutics*, 2010, 7 (4): 439–451.

Zonta M., Angulo M. C., Gobbo S., Rosengarten B., Hossmann K. A., Pozzan T., Carmignoto G., "Neuron-to-astrocyte signaling is central to the dynamic control of brain microcirculation," *Nat. Neurosci.*, 2003, 6 (1): 43–50.

Index